Ionic Liquids - Recent Advances

Edited by Pradip K. Bhowmik

Published in London, United Kingdom

Ionic Liquids - Recent Advances
http://dx.doi.org/10.5772/intechopen.1002097
Edited by Pradip K. Bhowmik

Contributors
Alfonso Martinez-Felipe, Ali Irfan, Aritri Biswas, Behrooz Maleki, Bhabani S. Mallik, Carlos Damián Rodríguez-Fernández, David King, Debalina Deb, Elena López Lago, Fatemeh Moosavi, Gamal A. Shazly, Haesook Han, Hong Guo, Huma Hameed, Jakob Smith, Junru Pang, Milad Ghani, Muhammad Zaman, Noreen Samad, Parisa Ghasemi Ilkhechi, Patricia Iglesias, Pouya Ghamari Kargar, Pradip K. Bhowmik, Raúl De la Fuente, Sadia Rani, Samaneh Sedigh Ashrafi, Samina Aslam, Seonghyeok L. Cox, Si L. Chen, Thamil Selvi Velayutham, Yago Arosa, Yousef A. Bin Jardan

First published in London, United Kingdom, 2025 by IntechOpen
IntechOpen is the global imprint of INTECHOPEN LIMITED, registered in England and Wales, registration number: 11086078, 167-169 Great Portland Street, London, W1W 5PF, United Kingdom

For EU product safety concerns: IN TECH d.o.o., Prolaz Marije Krucifikse Kozulić 3, 51000 Rijeka, Croatia, info@intechopen.com or visit our website at intechopen.com.

British Library Cataloguing-in-Publication Data
A catalogue record for this book is available from the British Library

Ionic Liquids - Recent Advances
Edited by Pradip K. Bhowmik
p. cm.
Print ISBN 978-0-85466-674-4
Online ISBN 978-0-85466-673-7
eBook (PDF) ISBN 978-0-85466-675-1

If disposing of this product, please recycle the paper responsibly.

Meet the editor

Pradip K. Bhowmik is currently a Professor of Chemistry and Biochemistry at the University of Nevada Las Vegas. Dr. Bhowmik has received several teaching awards, including the Regents' Teaching Award, the UNLV Foundation Distinguished Teacher Award, the UNLV Alumni Outstanding Faculty Member and the College of Sciences Distinguished Teacher Award. He received the POLY Fellow and ACS Fellow Award for his outstanding scientific and professional accomplishments and service to the American Chemical Society. He was honoured to be a Fellow of the Royal Society of Chemistry in the UK. Dr. Bhowmik has edited 14 books, authored nine book chapters, over 110 peer-reviewed journal articles, and is the inventor of five patents, and several patents are pending. He also sits on the Advisory Board of six companies.

Contents

Preface

Ionic liquids (ILs) are a unique class of materials composed of different cations and anions. They are called functional or task-specific materials that can be used in many applications in Chemistry, Physics, Biology, and Engineering. Their industrial applications are continuously increasing exponentially since they can fulfill the rigorous criteria because of their unique suite of properties, including their low volatility, high thermal stability, and ability to dissolve a wide variety of compounds and biopolymers, among others. However, this book discusses recent uses of dicationic ILs instead of monocationic as electrolytes for battery applications. It includes the polymerized IL gels as polymer electrolytes for supercapacitor applications. Chapters also include recent applications of ILs as lubricants and as pharmaceuticals for medicine. For their versatile applications, the physical properties, including microscopic dynamics, refractive index, activity coefficients, and molecular dynamics simulation of ILs, are also discussed in several chapters.

I hope this book will further stimulate research work in the ever-increasing field of ILs and, more importantly, highlight the need for an interdisciplinary approach, including the use of AI. Finally, I wish to thank the Editorial Office of the Book, particularly Iva Simcic Mance, for inviting me to edit the book and Dorian Salatic and Valentina Jolic for their help during the whole process.

Pradip K. Bhowmik
Department of Chemistry and Biochemistry,
University of Nevada Las Vegas,
Las Vegas, USA

Chapter 1

A Computational Study of Structure, Dynamics, and Spectral Properties of Ionic Liquids: Wavelet Spectra as a Tool to Assess Microscopic Dynamics of Ionic Liquids

Aritri Biswas and Bhabani S. Mallik

Abstract

This perspective summarizes the recent advances in modeling ionic liquids to highlight structural as well as dynamical aspects and spectroscopic investigations in complex ionic environments. Wavelet transform serves as a valuable tool for spectral analysis in these liquids. The wavelet spectra obtained from this transform are found to be compared with the results of the conventional Fourier methods. The merit of the wavelet transform method representing the vibrational stretch spectral signature depends on the flexible intermolecular potentials. Wavelet transform is an effective tool for calculating frequency fluctuations based on molecular dynamics simulations. Here, we emphasize the methodology of calculating wavelet spectra based on trajectories derived from computer simulations, aiding a molecular-level interpretation that correlates experimental results satisfactorily.

Keywords: ionic liquids, infrared probes, ultrafast spectroscopy, structural interactions, dynamical properties

1. Introduction

Ionic liquids (ILs) are molten salts consisting of ions held together by the attractive electrostatic force shown in **Figure 1(a)**, and they have a melting point of less than 100°C [1, 2]. The benefits of lowering the melting points of salts were found by incorporating large, asymmetric, flexible cations, and anions with delocalized charge distribution on them [3]. A combination of these IL properties inhibits crystallization and lowers the melting point, resulting in a new subclass of room-temperature liquid salts referred to as room-temperature ionic liquids (RTILs) [4, 5]. Also, ILs are hygroscopic, and it necessitates tuning the desired solvation properties of ILs for task-

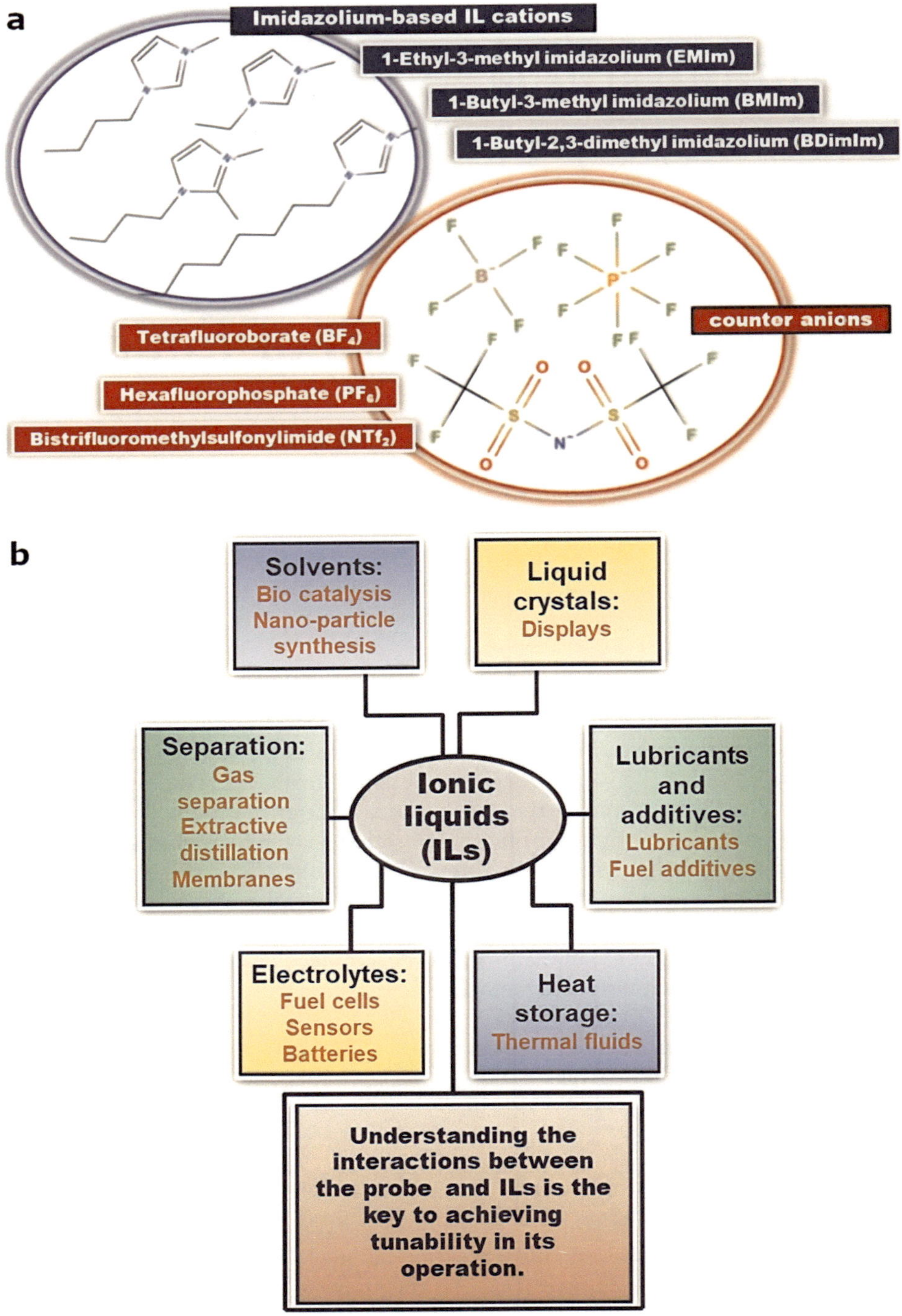

Figure 1.
(a) Ionic liquid (IL) constituting cations and anions used in research. (b) Various ionic liquid (IL) applications.

specific applications [2, 6]. Many scientists showed interest in using ILs in their respective fields after the successful synthesis of the air and water-stable ILs [7, 8]. Several of the IL properties such as low vapor pressure, solubility, and non-volatility make them accessible for use as green solvents [9, 10], and phase-transfer catalytic solvents [11–13]. ILs are chemically and thermally stable. Besides, owing to the low flammability, and a large electrochemical voltage window, ILs often find applicability as battery electrolytes [14, 15]. ILs also exhibit local charge ordering; the polar-apolar segregation maintains local charge neutrality. Hence, these liquid salts can solvate the hydrophilic and hydrophobic moieties and thus are attractive for separation. Nowadays, ILs also mitigate climatic changes and serve as a significant environmental concern [16]. They help eliminate the harmful, anthropogenic greenhouse gas, CO_2, by impinging the pores of organic and inorganic membranes called the supported ionic liquid membranes (SILMs), thereby facilitating carbon capture and storage (CCS) [17–19].

However, research on synthesis of ILs and classification of its properties have grown tremendously. A molecular-level picture of the interionic interactions, details of the ionic environment, and an atomistic interpretation would lead to a widespread application of these molecular liquids in several disciplines. Knowledge of the fundamental interactive forces operating within these molecular ions would help attain tunability in operation. We explore the unique structural variations and local dynamical transformations based on the spectral response of ionic solutions. Several experimental spectroscopic techniques [18–24] and theoretical studies [25–27] in the past years have determined the dynamics of structural reorganization in the ionic liquid framework. IR and Raman techniques were utilized to study the spectroscopic features, and assign bond stretching and bending vibrations provided timescales are slow. However, ultrafast nonlinear spectral measurements [28, 29] capture ultrafast processes occurring on sub-picosecond timescales. The time-resolved spectral technique led to the assignment of spectral peaks and bands and helps explore chemical exchange reactions [30], hydrogen bond network evolution [31–33], water dynamics [20, 34], dynamics of internal rotation [35, 36], vibrational population transfer [37], as well as protein structure analysis [38–41].

Infrared (IR) probes interrogate structural dynamics in ILs to capture the response of the liquid microenvironment. Spectral profiles are primarily governed by the subtle interplay of the long-ranged electrostatic attractive Coulomb force, short-ranged dispersive forces, and the strong-directional hydrogen-bonding interactions. These delicately balanced interactive forces result in structural and dynamical complexity, that is, heterogeneous structure leading to heterogeneous dynamics. Probes considered are either the external probes such as the pseudo-halide salts, negatively charged ions like SCN^- [24, 25], $SeCN^-$ [4, 5], positively charged species such as 2-SeCN-$BMIm^+$ [17], and neutral moieties like D_2O, H_2O, CO_2 [22, 23], or the intrinsic vibrational modes of interest such as the C-D vibrational stretching mode at the C_2 position of imidazolium cation [42] that can directly or indirectly access the microscopic heterogeneous IL dynamics. The criteria for selecting the probe mode for experimental studies depends on—(i) long vibrational time window that enables the examination of dynamics over a substantial timeframe, (ii) small size of the vibrational probe molecule so that they do not interact to affect the overall liquid structure, and (iii) higher extinction coefficient such that the probe modes strongly absorb at a particular wavelength. Moreover, the Stark effect further governs spectral dynamics through the electric field coupling with the molecular dipole. While heteronuclear diatomic molecules like SCN^-, $SeCN^-$ with a permanent dipole moment couple with the corresponding electric field vector through first-order Stark effect [43], the homonuclear diatomic probe molecule such as CO_2 couples is through a more complex second-order Stark effect [44, 45]. The frequency of the vibrational chromophores or infrared probes is sensitive to the local environment. At an initial instant of time, the probe modes oscillate at a particular intrinsic frequency. With time, differential site-specific interactions of the probe with the surrounding ionic environment result in vibrational frequency changes. The randomization of the probe's vibrational frequency is referred to as vibrational spectral diffusion [46–49], and the infrared probes monitor spectral diffusion dynamics by analyzing the structural and dynamical response of the IL constituting cations and anions.

In the past years, the microscopic picture of spectral diffusion dynamics and hydrogen bond breaking and reformation dynamics of the OD stretch local modes of deuterated water, [48] was investigated employing first-principles molecular dynamics (FPMD) [50, 51] simulations in conjunction with the time-series wavelet transform [52, 53]. Also, a vast majority of previous studies focused on the ionic hydration shells of F^- [54], Cl^- [55], and Br^- [56]. Such studies [54–56] investigated the dynamical response of solvent molecules in halide ion-water systems, empirically correlating the vibrational frequencies to the instantaneous configurations of the proximal solvent environment. Spectral diffusion dynamics inside the ion-water hydration shells [54–56] reveals three timescales: a fast and rapid initial decay short-time oscillation component associated with the dynamics of intact hydrogen bonds, followed by an intermediate relaxation time constant corresponding to the lifetime of hydrogen bonds and the longer time decay component representing the escape of solvent molecules from the ion hydration shell. The structural, dynamical, and spectral properties of the supercritical fluids [57, 58] were also explored to pinpoint the differences in supercritical solutions compared to those at ambient temperatures [48, 55]. The supercritical systems [57, 58] in the past proceedings were observed to exhibit faster spectral diffusion dynamics and hydrogen bond kinetics than the corresponding solutions under ambient conditions [48, 55]. Biswas and Mallik further carried out density functional theory (DFT)-based FPMD simulation with the van der Waals (vdW) dispersion correction to inspect the dynamics of solute-solvent hydrogen bonds and vibrational spectral signature of solvent water in aqueous solvation of amide [59]. This study unveils the asymmetric solvation behavior based on the hydrophilic hydrogens attached to amine nitrogen and hydrophobic interactions of the hydrogen atom attached to carbon in the simplest amide molecule [59]. Moreover, the first principle theoretical study in deuterated aqueous methylamine solution yielded valuable insights on solvation structure, associated dynamical properties, hydrogen-bonding interactions, exchange dynamics, and time-dependent vibrational spectral dynamics [60]. Furthermore, spectroscopic calculations were executed in pure methylamine systems [61]. Moreover, the extent of hydrophobicity of amides, N-methyl formamide (NMF), N,N-dimethylacetamide (DMA), and N-methyl acetamide (NMA) affecting the solvent dynamics as well as the structure and stretching dynamics around the tertiary amphiphilic amide, N, N-dimethylformamide (DMF) were highlighted [62, 63]. In an earlier report, the structural dynamics of the amide bands were also probed to understand the solvent coupled dynamics and vibrational stretching profiles of the C═O, N—H, and C—N modes of amide derivatives in water [64]. Biswas and Mallik further performed QM/MM molecular dynamics simulations of N-methyl acetamide in different polar and non-polar solvents to track the spectral response of solvating water molecules based on polarity and hydrogen-bonding ability of the surrounding solvent environment [65]. Besides, the effects of a weakly coordinating anion, tetrafluoroborate ($BF4^-$), were inspected on nearby solvent molecules [66], and the heterogeneous occupancy and vibrational dynamics of spatially patterned water molecules in the vicinity of aqueous divalent anions, sulfate, and thiosulfate with varied local density [67] were summarized. Further, the solvation nature of hydrophobes and spectral dynamics of water molecules near the hydrophobic molecular solutes like the tetramethylammonium (TMA) cation [49], neopentane (NEO), and tetra-butyl alcohol (TBA) [68] were also studied. Moreover, several of the past studies explored the distinctive behavior of water molecules inside the hydration shell of glycine-water system [69], aqueous acetone solution at room temperature [70], and the conformation-induced spectral response of proximal solvent water inside the hydrogen peroxide solvation shell [71]. But, the spectral diffusion timescales in the solvation shell of aqueous hydrogen peroxide [71] were distinct from the hydration shell of ionic solutes and neutral solutions [49, 63]. The escape dynamics of water molecules from the H_2O_2 solvation shell were probably missing because of the rigidity of the hydration shell. Perhaps, the demarcation between the solvation shell and bulk water got blurred owing to no significant change in the average stretching frequencies of the hydrogen-bonded hydroxyls inside the hydration shell and bulk water molecules. Hence, the escape events were not noticeable inside the H_2O_2-water solvation shell.

Moreover, all these aforementioned studies deploy ab initio molecular dynamics (AIMD) simulations [50, 51] for trajectory generation and wavelet-based frequency calculations [52, 53] for determining the equilibrium distribution. However, the calculated average stretch frequency and the computed peak position of maximum intensity obtained from the AIMD simulations using the dispersion-corrected DFT functionals exhibit a slight deviation from the corresponding experimental value. The shift in the calculated vibrational stretch spectral signature compared to the experiment is likely due to the systematic errors introduced in the present calculations using different density functionals, finite basis set cut-offs, fictitious electronic mass, and other simulation parameters of the electronic structure calculations [72, 73]. Most likely, the smaller system size resulting in the spurious correlation of ions is often associated with the finite-size and time effects [74–78]. However, choosing a relatively smaller system size may be one of the viable options due to the significant computational cost of calculations employing FPMD simulations. Moreover, the observed results of the FPMD simulations with different density functionals are not much affected by the finite-size and finite-time effects. However, systematic errors could be efficiently dealt with using a constant scaling factor [79, 80]. But the primary objective in these ab initio computational studies was to determine the dynamics of vibrational spectral diffusion, and the normalized dynamics would remain unaltered by the discrepancies in the computed frequencies. So, such a scaling factor was not introduced in the calculations to bring a closer agreement between the calculated average vibrational stretching frequency and the experiment. Despite all of these, to capture the realistic probe response, one necessarily needs to consider a large system with a sufficient number of probe modes to obtain statistically meaningful results. So, we should either perform long enough simulations of a solute probe immersed in the proximal solvent environment or consider many probe molecules to get better statistical averaging. However, neither of these options could be possible while executing FPMD simulations. So, we had performed the empirical potential-based classical molecular dynamics (MD) simulations and had applied the time-series wavelet transform (WT) for the first time for ILs to study the ultrafast aqueous dynamics of the deuterated hydroxyl probe modes in the water-IL mixtures [81].

2. Wavelet transform of classical trajectories (WTCT) an indispensable tool for spectroscopic analysis of ILs

Transform in terms of mathematical methods like Fourier, Hilbert, or wavelet is an appropriate technique to obtain simplified signals from a function of a few variables. Fourier transform (FT) is the first method to provide the signal by applying filtering and convolution/deconvolution process. Moreover, this analysis is adopted to analyze the data in the frequency domain but not possible in the time domain. One needs to localize the signal in both time and frequency domains for time-dependent variables. WT, a mathematical tool, is very useful for spectral analysis with analyzing functions called wavelet with localization properties in both frequency and time domains [82]. Within this method, a signal is decomposed into a set of basic functions, known as wavelets, which are small waves that appear in different time intervals. When FT is unable to provide information in both time and frequency domains, WT is very useful in overcoming the disadvantages of FT and eliminating the resolution issue in the analysis of signals. WT resembles a power spectrum locally in space, providing spectral properties of the signal at different locations in physical space [83]. Moreover, Perrier et al. derived the analytical relation linking the mean wavelet spectrum to the classical Fourier spectrum; they concluded that the wavelet spectrum obtained from WT was mathematically proved to appear as an average of the Fourier spectrum weighted by the square of the Fourier transform of the wavelet [83]. Although the wavelet spectrum is biased by a particular wavelet used in the transform process, accurate results can be obtained by choosing the appropriate mathematical form of wavelet and performing the backtesting for a specific purpose. Moreover, high-frequency near-field sounding, IR-spectroscopy, NMR experiments, and model harmonic signal were used to conclude that the WT provides us an adequate description of the dynamics of quasi-harmonic processes that take place in the system having non-stationary states. Indeed, wavelet spectra could be compared with the results of the conventional Fourier methods [84]. Due to these reasons, the WT has been gaining popularity in various scientific and engineering fields.

The time-series wavelet method determines the fluctuating frequencies of the vibrational probe mode of interest based on variable functions of the associated chemical bond. Past studies reported the Laskar [85, 86] approach and the Arevalo and Wiggins [53] method of determining the instantaneous vibrational stretch frequency based on molecular dynamics trajectories. While windowed Fourier transforms on short, disjoint time intervals of the entire trajectory timeframe result in the generation of the vibrational stretching frequencies following the Laskar methodology. The method of Arevalo and Wiggins calculates the time-dependent frequencies *via* continuous WT of the trajectories. However, the time-series wavelet transform of Arevalo and Wiggins [53] is the preferred method of choice to determine the instantaneous vibrational frequencies of the probe molecule in the complex molecular environment as well as calculate the normalized stretch frequency distribution in all previous reports [48, 49, 55, 60, 61, 81]. WT is one of the most effective mathematical tools that automatically adjusts the size of the time window to the oscillation frequency, localizing time in a better way and providing detailed spectral information than the Laskar method of the windowed Fourier transforms. In the wavelet method of Arevalo and Wiggins [53], the instantaneous frequencies are generated at each time interval through Fourier transformation [87] of the time domain data at each timestep. Further, the WT also accounts for the intermodal coupling effects in frequency calculations [48, 55]. The scheme for the wavelet transform is shown in **Figure 2**.

WT initiates with the construction of a complex time-dependent function $f(t)$ consisting of the real and imaginary components, $r(t)$ and $ip(t)$, respectively:

$$f(t) = r(t) + ip(t) \tag{1}$$

In the above expression, the instantaneous distance fluctuations along the unit vector are denoted by the real term, $r(t)$ and the momentum along the specific bond is expressed by the imaginary constituent, $ip(t)$.

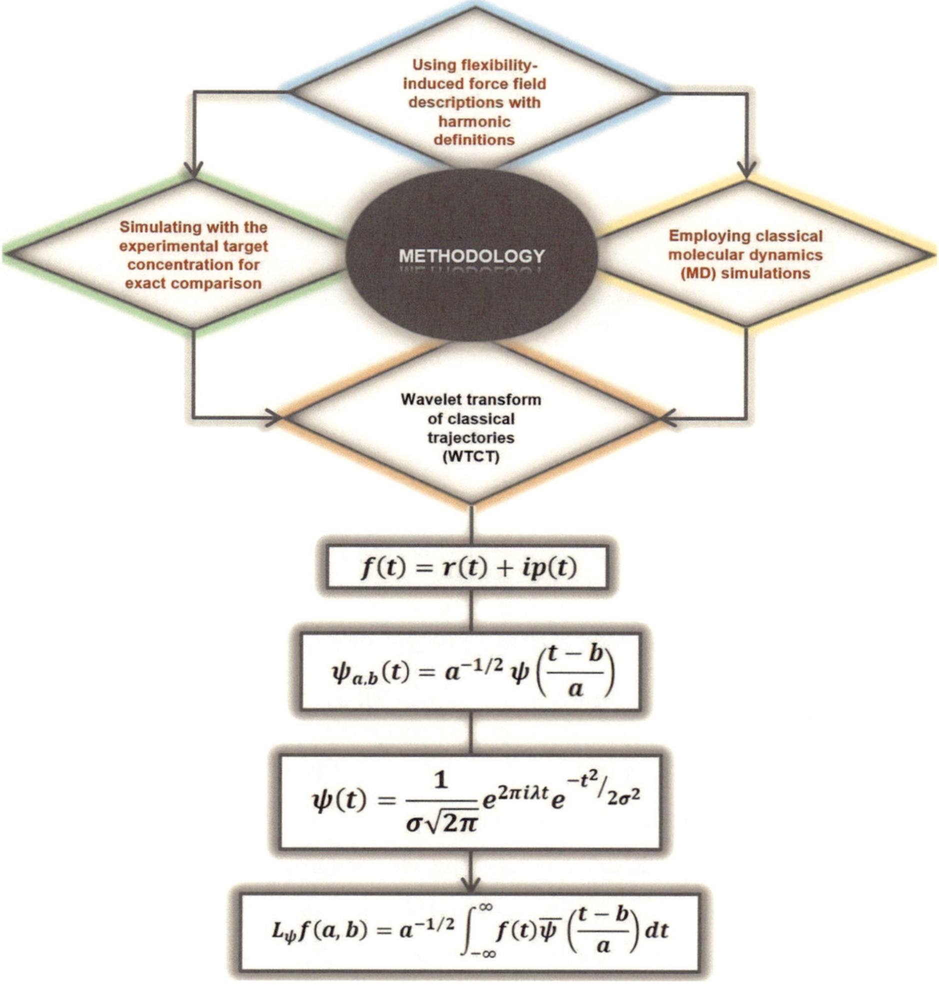

Figure 2.
Schematic of theoretical methodology, wavelet transform of classical trajectories (WTCT) employed to calculate the vibrational stretching frequency of the probe mode of interest.

Next ψ represented in terms of two variables, the scale and time variable a, b respectively, expresses the translations and dilations of the mother wavelet ψ

$$\psi_{a,b}(t) = a^{-1/2}\psi\left(\frac{t-b}{a}\right) \quad (2)$$

The criterion for choosing the a and b values of $\psi a,b(t)$ is such that ψ is real, $\psi a,b$ (t) decays to 0 at $t \to \pm\infty$ and the admissibility condition, $0 < c_\psi = 2\pi \int_{-\infty}^{+\infty}\left(|\hat{\psi}(\omega)|^2/|\omega|\right) d\omega < \infty$, is satisfied in which $\hat{\psi}$ represents the Fourier transform of ψ. Next,

$$L_\psi f(a,b) = a^{-1/2}\int_{-\infty}^{\infty} f(t)\overline{\psi}\left(\frac{t-b}{a}\right) dt \quad (3)$$

Wherein, $a > 0$, and b is real.

Time-frequency representations at each time interval $t = b$ stores local information in $L_\psi f\ (a, b)$.

Now, we analyze using the Morlet-Grossman wavelet [88].

$$\psi(t) = \frac{1}{\sigma\sqrt{2\pi}} e^{2\pi i\lambda t} e^{-t^2/2\sigma^2} \quad (4)$$

In the above equation, the shape parameter σ^2 controls the shape of the Gaussian time window, and the center frequency λ could be readily varied. Mostly, the values of λ and σ considered were 1 and 2, respectively, and those values could be tuned to get a better resolution in the calculated frequencies.

In our calculations, we considered a periodic oscillatory function, $f(t) = e^{i2\pi\omega_0 t}$ and the WT of $f(t)$ employing Eq. (3) yields

$$\begin{aligned} L_\psi f(a,b) &= a^{-1/2}\int_{-\infty}^{\infty} e^{i2\pi\omega_0 t}\overline{\psi}\left(\frac{t-b}{a}\right) dt \\ &= a^{1/2}\int_{-\infty}^{\infty} e^{i2\pi\omega_0 as} e^{i2\pi\omega_0 b}\overline{\psi}(s)ds \\ &= a^{1/2} e^{i2\pi\omega_0 b}\overline{\psi}(\omega_0 a) \end{aligned} \quad (5)$$

Wherein, the modulus of WT $L\psi f(a, b)$ is given by

$$|L\psi f(a,b)| = \sqrt{a}\overline{\psi}(\omega_0 a) \quad (6)$$

In the above expression, $\overline{\psi}(\omega) = e^{-2\pi^2\sigma^2(\omega-\lambda)^2}$ and $\overline{\psi}$ provides for localization in frequency with the global maximum lying at $a\omega_0 = (1/2)\left(\lambda + \sqrt{\lambda^2 + (1/2\pi^2\sigma^2)}\right)$. Thus, the frequency variable defined with the Morlet-Grossman wavelet is as follows

$$\omega(t=b) = \frac{1}{2a}\left(\lambda + \sqrt{\lambda^2 + \frac{1}{2\pi^2\sigma^2}}\right) \quad (7)$$

We calculate the instantaneous vibrational stretching frequency at each time interval $t = b$, from the inverse of the scale factor a. The scale value a governs the

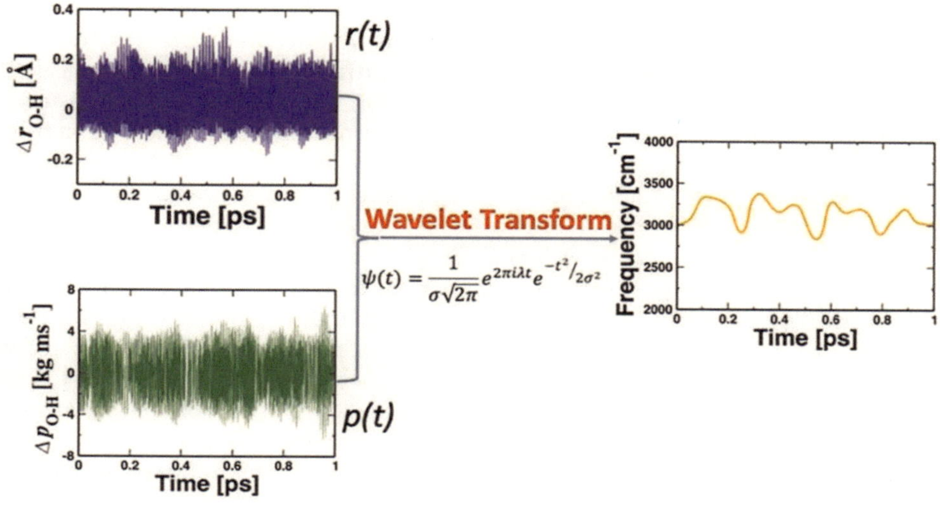

Figure 3.
A flowchart representing the input variables (fluctuations in the distance, r(t) and relative momentum, p(t)) in the time domain to determine the instantaneous vibrational stretching frequencies of a particular representative probe, OH covalent bond, in the time domain applying the WT of classical trajectories (WTCT).

width of the time window. At high frequency, the small ***a*** value automatically narrows the time window and widens it at a lower frequency due to the large ***a*** value. Wavelet transform reruns through the entire trajectory timeframe to determine the time-dependent fluctuating frequencies and frequency distribution of the vibrational probes comparable to the experimental absorbance spectrum.

We now summarize the calculation of the frequency stretch distribution of the vibrational probes step-by-step in a detailed manner, employing wavelet. Firstly, we calculate the distance of the unit bond vector and then, we determine the average of the intramolecular bond distance. Next, we estimate the time-varying instantaneous distance fluctuations of the probes. Then, we compute the relative velocities and velocity projections along a specific bond. Henceforth, we further determine the momentum along the specific bond vector, multiplying the reduced mass of the atoms involved with the relative velocities. A schematic presentation of these variables is shown in **Figure 3**. Thereafter, the variables ***a*1**, ***a*2** corresponding to the largest modulus of WT are determined. Further, Simpson's formula for numerical integration estimates the modulus of the WT at scale value ***a*** and time value ***b***. Subsequently, the scale ***a*** corresponding to the maximum is isolated, and the inverse of the scale value ***a*** yields frequency about time ***b***. Further, the WT repeats itself through the entire trajectory over the total simulation time, and the frequencies are obtained. Past studies revealed that the wavelet theorem had already been applied to determine spectral diffusion dynamics in neat systems [48, 89], aqueous solvation shells [54, 55], solution of neutral moieties [49, 59, 90], and ionic liquid solutions [81].

3. Water dynamics in aqueous ionic liquid solutions

Aqueous ionic liquid-based electrolytes have been attractive owing to their diverse use in chemical sciences shown in **Figure 1(b)**. However, we need to fully understand the properties of these ILs to enhance their applicability further. Almost all ILs are hygroscopic; thus, it is crucial to know how the smallest possible water concentration interacts with the surrounding ionic environment, affecting the physicochemical properties of the aqueous ionic solutions. Most often, water molecules tend to accelerate the ionic liquid dynamics at higher water concentrations. Our study primarily focused on monitoring the probe response in aqueous ionic solutions, lithium bis (trifluoromethylsulfonyl) imide ($LiNTf_2$) and 1-ethyl-3-methyl imidazolium bis (trifluoromethylsulfonyl) imide ([EMIm][NTf_2]) to examine the ion-probe interactions in the presence of water as well as its impact on the spectral properties.

The reason for taking up these systems with the identical anionic constituent, bis (trifluoromethylsulfonyl) imide or bistriflimide ions and the lithium versus the heterocyclic imidazolium cation are to investigate the cation-dominated changes in frequency profile as well as the spectral dynamics of the vibrational probe modes in the aqueous ionic liquid framework. An aqueous $LiNTf_2$ solution at room temperature is used as battery electrolytes [91] due to its large electrochemical time window.

Besides, [EMIm][NTf_2] is chosen for this study among other imidazolium-based ILs due to its short cationic chain length that inhibits the segregation of the polar-apolar domains. Moreover, [EMIm][NTf_2] is a less viscous IL, both air and water-stable, with the potential for battery applications [92, 93] owing to its good thermal and electrochemical stability and low flammability. Recently, imidazolium-based ILs have also been used for CCS [94] as they preferentially dissolve the greenhouse gas CO_2.

The simulated systems analyzed were 5% HOD-H_2O in $LiNTf_2$ and [EMIm][NTf_2] at a room temperature of 300 K, consisting of 100 HOD probe molecules. Experimentally, at room temperature, it was difficult to achieve the water-to-IL ion-pair ratio comparable in these two solutions due to the unavoidable solubility conditions [20]. So, we simulated both these systems according to the experimental target concentration, [20] 2.5:1 $H_2O/LiNTf_2$ and 0.67:1 H_2O/[EMIm][NTf_2]. Initially, the concentration discrepancies might be thought to be responsible for the structural and dynamical differences existing in the solutions. However, the chosen concentrations were found reasonable enough to deduce the probe response owing to a slight change in the cationic constituent of the water-IL mixtures.

3.1 Vibrational stretch spectral signature in the normalized frequency distribution

We explore the dynamical response of the probe OD modes of HOD molecule in the solvated ionic liquid environment. The deuterated hydroxyls or the O-D stretch modes serve as a sensitive reporter of the aqueous system dynamics. **Figure 4(a)** represents a comparative plot of our calculated normalized stretch frequency distribution of the OD bonds of the probe applying the WTCT method versus the experimental linear absorption spectral results [20].

The linear absorbance spectrum of HOD in bulk water molecules in the previous proceedings [20, 96] displayed a broad, inhomogeneous spectral lineshape with its peak position located at 2509 cm^{-1}, representing various hydrogen-bonding environments [97]. The calculated average vibrational frequency of the OD stretch modes and the corresponding position of maximum peak intensity in aqueous $LiNTf_2$ and [EMIm][NTf_2] solutions are 2636 and 2645 cm^{-1}, respectively. Thus, the addition of ionic liquid constituent ions led to frequency shifts toward the blue side of the spectrum. Based on past reports, the probe modes were completely separated from each

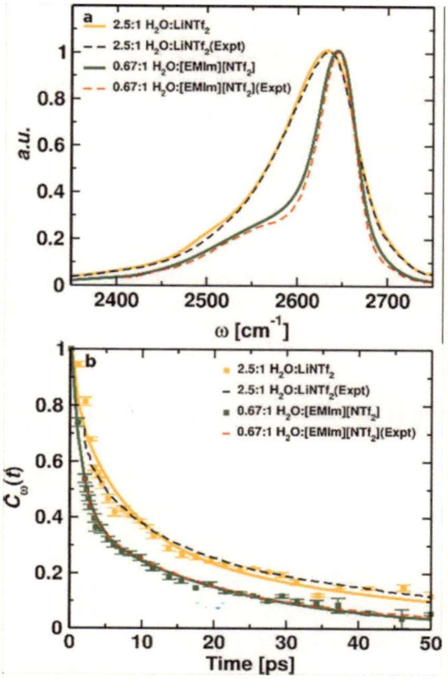

Figure 4.
Comparative spectral analysis of the deuteroxyl probe modes in an aqueous solution of LiNTf2 versus [EMIm][NTf2]. (a) The solid lines represent the normalized vibrational stretch frequency profile of the water-IL mixtures, and the dotted lines denote the experimental linear FT IR spectral results [20]. (b) The time-dependent decay of the frequency-frequency correlation function (FFCF) of the probe OD modes. At the same time, the squares depict the calculated data. The solid curves represent the fitted data, fit by a tri-exponential fitting function. Here too, the dotted lines represent the experimental results. The error bars provide an estimation of the standard deviation within the simulated datasets. The comparatively plotted data sets obtained from experiment [20] were abstracted using Origin version 9.60 [95]. Copyright 2020 American Chemical Society.

other in dilute aqueous solutions. The probe-probe interactions were almost negligible at a lower water concentration, which resulted in a narrower distribution with a blue-shifted spectral peak representing ion-probe interactions [96, 98]. But, with an increase in water concentration, water molecules tend to associate with the nearby solvent water in the aqueous ionic solutions, forming water-water hydrogen-bonding networks that give rise to a second population in the vibrational stretch spectrum [99]. Thus, the lineshape of the spectral signature broadens at a higher water concentration, exhibiting a bimodal lineshape. The broad, asymmetric lineshape represents a principal spectral peak associated with the ion-bound water population along with a smaller hump at a lower wavenumber representing water-water hydrogen bond associations. It is further observed that the ion-associated deuterated hydroxyl probe modes are weaker than the water-water hydrogen-bonded configurations in aqueous ionic solutions [100, 101].

For both the aqueous mixtures, the peak positions of maximum intensity in the computed frequency distribution correlate with the experimentally reported peak center frequency, 2635 cm^{-1} for $LiNTf_2$ and 2645 cm^{-1} for the [EMIm][NTf_2] solution. Besides, the full width at half maxima (FWHM) of the spectral peak in the computed distribution was 78 cm^{-1} for $H_2O/LiNTf_2$ and 46 cm^{-1} for H_2O/[EMIm][NTf_2]. The width of the spectral lineshape in the calculated equilibrium distribution was similar to the experimentally recorded FWHM, 78 cm^{-1} in the lithium salt and

44 cm^{-1} in the imidazolium-based IL solution. A blue shift of 9 cm^{-1} was observed on substituting the lithium cation with the heterocyclic imidazolium ion, reflecting that the ion-water hydrogen bonds were weaker in the [EMIm][NTf_2] solution than in the aqueous lithium salt solution. Both the ion-water and the water-water hydrogen bond populations were noticeable in the normalized frequency distribution of the H_2O/[EMIm][NTf_2] solution, whereas the distinction between the two populations was not clearly defined in the distribution of the aqueous lithium mixture. In the salt solution of the lithium cation, the two populations substantially overlapped, resulting in a broader continuum without any distinctive water-associated spectral peak at around ~2550 cm^{-1}. We observe an asymmetric bimodal shape in the vibrational stretch signature of the [EMIm][NTf_2] solution.

3.2 Dynamics of vibrational spectral diffusion: Frequency-time correlations

To gain insight on water dynamics associated with the ultrafast molecular motions, we compute the time-dependent decay of frequency fluctuations, that is, the decaying dynamics of the frequency-frequency correlation functions (FFCFs), $C_\omega(t)$. We used the instantaneous fluctuating vibrational stretching frequencies obtained from the time-series wavelet analysis to calculate the frequency autocorrelation functions. At time t = 0, the local vibrator oscillates at a particular frequency value. However, as the vicinal environment surrounding the probe changes with time, the vibrational frequency fluctuates such that the vibrational oscillator acquires a different frequency at time t. Thus, the joint probability of the frequency-time correlations averaged over all initial frequencies providing an estimate of vibrational spectral diffusion [48, 71, 102] is given by the following equation

$$C_\omega\ (t) = \frac{\langle \delta\omega\,(t)\,\delta\omega\,(0)\rangle}{\langle \delta\omega\,(0)\,2\rangle} \tag{8}$$

where $\delta\omega(t)$ refers to the instantaneous frequency shift from the average frequency value. Tri-exponential fits the correlation function decay curves using Eq. (9) yields the dynamical timescales associated with the fluctuating frequencies of the OD bonds in the water-IL mixtures:

$$f(t) = a_1 e^{-t/\tau_1} + a_2 e^{-t/\tau_2} + a_3 e^{-t/\tau_3} \tag{9}$$

Figure 4(b) depicts FFCF decay curves representing the dynamics of vibrational spectral diffusion. The correlation function decay yields the dynamical timescales associated with the fluctuating frequencies of the OD bonds in the aqueous IL environment. The time-dependent decay of frequency fluctuations in the water-IL mixtures portrays a rapid initial decay time constant indicating structural fluctuations of the local vibrational probes, followed by an intermediate timescale depicting larger length-scale fluctuations of the hydrogen-bonded configurations [25, 98] and the slowest decay time representing the exchange of the hydrogen-bonded networks *via* angular jumps [103, 104]. While the decay timescales in 2.5:1 water/$LiNTf_2$ solution are 0.85, 7.21, and 41.82 ps, the reported time constants in the [EMIm][NTf_2] mixture are 0.76, 1.85, and 20.98 ps, well in agreement with the ultrafast experiment [20]. The faster timescale of the FFCF decays is almost similar in the two solutions. However, the intermediate and the longer decay time constants gradually lengthen in the aqueous lithium salt solution compared to the imidazolium-IL solution. Hence, even though more water molecules in the $LiNTf_2$

solution are supposed to speed up the IL dynamics. The spectral diffusion timescales in the aqueous ionic liquid mixtures show a dynamical trend that is opposite to what was expected. Instead of the lithium salt solution, water molecules in the [EMIm][NTf_2] solution exhibit faster dynamics. The cation-dominated spectral response also indicates the changes in local molecular structure and ultrafast dynamical transformations induced by these structural modulations in both the water-IL mixtures. Extra water molecules in an aqueous lithium solution facilitate the solubilization of the highly crystalline salt and are engaged in the molecular arrangement of the restricted ionic environment. In contrast, the water in the water-[EMIm][NTf_2] solution accelerates the overall system dynamics despite no additional water molecules. Hence, depending on the characteristic behavior of water molecules in the aqueous ionic environment, water plays a specific role in controlling dynamics in these ionic liquid solutions.

4. Conclusions and outlook

The success of any approach in reproducing the realistic picture depends on the proper methods employed to calculate the properties and the choice of a computationally feasible model that yields accurate results in optimum time. Molecular dynamics simulation [105] of ILs poses some interesting new challenges; it was difficult to model electrostatic forces and Lennard-Jones potentials; high viscosity expands the simulation period. So, obtaining fully equilibrated IL systems with a reasonably large simulation size was difficult. The other challenges were to define a simulation environment that enables cost-effective computation and can accurately predict the experimental results, thereby linking simulations to experiments. The molecular modeling combined with an appropriate spectroscopic tool like WT helps find the detailed interpretation of the IL properties. Experimentally, this field has grown tremendously, but there is a scope from the computational perspective despite the available studies based on a single vibrational probe molecule [25, 26, 106] with a substantial computational cost [107, 108]. Past studies [109, 110] also declared the requirement of an electronically polarizable model to reproduce the experimental results correctly. However, polarizable force fields in ionic liquid simulations incur substantial computational costs. Further, the ionic liquid simulations pose unprecedented challenges due to the anisotropy of the constituting molecular ions, the finite-size effect due to the relatively large size of ions, and the segregation of the polar-apolar domains in the IL systems [105]. Thus, exploring the links between molecular architecture, ultrafast dynamics, and spectral response characterizing cation-anion interactions is necessary to understand the heterogeneous dynamics of ILs from spectral profiles. Work on theoretical ultrafast spectroscopy assessing spectral diffusion dynamics on ionic liquids primarily highlights the study of ionic liquid properties based on spectral response. Experiments often led to the characterization of new physicochemical IL properties. However, the tuning of those properties toward desired uses necessitates an understanding of the molecular picture of the intricate IL environments and an atomistic interpretation of the interionic interactions, and interactions with the surrounding solvent medium as well. Herein, we looked at the hygroscopic IL behavior and the impact of water or aqueous environment on the ultrafast IL dynamics. Based on the local response of the inbuilt or intrinsic and extrinsic vibrational probes in these molten liquids, we can predict structural variations, as well as dynamical interactions occurring at ultrafast timescales that would aid in tuning IL properties for specific desired purposes.

Further, the need was to define a simulation environment that enables cost-effective computation and correlates the experimental results. WTCT method involves calculations of wavelet spectra of probes in ILs simulated employing classical force-field-based molecular dynamics with flexibility-induced intermolecular interaction parameters based on harmonic descriptions. We implemented the WTCT to calculate the instantaneous vibrational stretching frequencies of the probe mode of interest and compute the wavelet spectra [52, 53]. The wavelet-based frequency calculations applied for the first time for ionic liquids determine the fluctuating frequencies and explore the phenomena of vibrational spectral diffusion [48, 49, 61] to aid in the molecular-level interpretation of the spectral response. Here, we demonstrate the application of the WT method in determining the spectral properties of the probe stretch modes to sensitively report the local structural fluctuations and ultrafast dynamics of the surrounding ionic environment. Further, we correlate the observed results by adopting this methodology with the available experimental data, validating our computational method.

A comparative analysis of the normalized vibrational stretch frequency distribution of the probe mode and the experimental absorbance spectra provides a satisfactory performance of the employed method. The wavelet method employed in the current work computes the frequencies, scanning over a broad frequency range. One would get information about the principal spectral peak and the low- and high-frequency spectral features of the normalized distribution applying the WT. This method can selectively isolate a specific probe mode from the complicated molecular arrangements. It also provides preliminary information about the vibrational stretch spectral signature and the spectral processes in complex systems. Also, this methodology can be adapted to calculate the instantaneous stretching frequency of any probe of interest in a complex molecular environment, predict the properties, and attain detailed information a priori about intricate systems wherein obtaining the spectral data is very difficult and sensitive to experiments. Further, the wavelet spectral study of the infrared probes in ILs [111–115] based on this methodology provided a satisfactory model of the theoretical vibrational spectra, bridging the existing gap between theory and experiments. Further, the methodology could also be extended to the solvation of battery electrolytes [116, 117], interfaces [118–121], micelles [122], nanoparticles [123–125], peptides, polypeptides [126, 127], and phospholipid multilayers [128, 129], non-aqueous systems [32, 130, 131], and biomolecules [30, 132]. Thus, the adopted simulation methodology helps overcome the challenges encountered in IL modeling. All in all, the theoretical methodology based on the wavelet transform of trajectories is a cost-effective, and convenient way that give us an idea of the heterogeneous environment of IL systems, providing a realistic molecular-level picture comparable to experiments.

Acknowledgements

Aritri Biswas likes to thank the Ministry of Education, India, for her Ph.D. fellowship.

Author details

Aritri Biswas* and Bhabani S. Mallik
Department of Chemistry, Indian Institute of Technology Hyderabad, Sangareddy, Telangana, India

*Address all correspondence to: aritri.payel@gmail.com

References

[1] Walden P. Molecular weights and elect conductivity of several fused salts. Bulletin of the Russian Academy of Sciences. 1914;**8**:405-422

[2] Welton T. Room-temperature ionic liquids. Solvents for synthesis and catalysis. Chemical Reviews. 1999;**99**(8): 2071-2084

[3] Hurley FH, Wier TP. Electrodeposition of metals from fused quaternary ammonium salts. Journal of the Electrochemical Society. 1951;**98**(5): 203-206

[4] Tamimi A, Bailey HE, Fayer MD. Alkyl chain length dependence of the dynamics and structure in the ionic regions of room-temperature ionic liquids. The Journal of Physical Chemistry. B. 2016;**120**(30):7488-7501

[5] Tamimi A, Fayer MD. Ionic liquid dynamics measured with 2D IR and IR pump–probe experiments on a linear anion and the influence of potassium cations. The Journal of Physical Chemistry. B. 2016;**120**(26):5842-5854

[6] Hallett JP, Welton T. Room-temperature ionic liquids: Solvents for synthesis and catalysis. 2. Chemical Reviews. 2011;**111**(5):3508-3576

[7] Wilkes JS. A short history of ionic liquids—From molten salts to neoteric solvents. Green Chemistry. 2002;**4**(2): 73-80

[8] Wilkes JS, Zaworotko MJ. Air and water stable 1-ethyl-3-methylimidazolium based ionic liquids. Journal of the Chemical Society, Chemical Communications. 1992;**13**: 965-967

[9] Ueno K, Tokuda H, Watanabe M. Ionicity in ionic liquids: Correlation with ionic structure and physicochemical properties. Physical Chemistry Chemical Physics. 2010;**12**(8):1649-1658

[10] Castner EW, Margulis CJ, Maroncelli M, Wishart JF. Ionic liquids: Structure and photochemical reactions. Annual Review of Physical Chemistry. 2011;**62**:85-105

[11] Bender J, Jepkens D, Hüsken H. Ionic liquids as phase-transfer catalysts: Etherification reaction of 1-octanol with 1-chlorobutane. Organic Process Research and Development. 2010;**14**(3):716-721

[12] Kumar V, Talisman IJ, Bukhari O, Razzaghy J, Malhotra SV. Dual role of ionic liquids as phase transfer catalyst and solvent for glycosidation reactions. RSC Advances. 2011;**1**(9):1721-1727

[13] Ross J, Chen W, Xu L, Xiao J. Ligand effects in palladium-catalyzed allylic alkylation in ionic liquids. Organometallics. 2001;**20**(1):138-142

[14] Tigelaar DM, Meador MAB, Bennett WR. Composite electrolytes for lithium batteries: Ionic liquids in APTES cross-linked polymers. Macromolecules. 2007;**40**(12):4159-4164

[15] Lewandowski A, Świderska-Mocek A. ChemInform abstract: Ionic liquids as electrolytes for li-ion batteries: An overview of electrochemical studies. Journal of Power Sources. 2009;**194**: 601-609

[16] Yamada SA, Bailey HE, Tamimi A, Li C, Fayer MD. Dynamics in a room-temperature ionic liquid from the cation perspective: 2D IR vibrational echo spectroscopy. Journal of the American Chemical Society. 2017;**139**(6): 2408-2420

[17] Shin JY, Yamada SA, Fayer MD. Dynamics of a room temperature ionic liquid in supported ionic liquid membranes vs the bulk liquid: 2D IR and polarized IR pump–probe experiments. Journal of the American Chemical Society. 2017;**139**(1):311-323

[18] Shin JY, Yamada SA, Fayer MD. Carbon dioxide in a supported ionic liquid membrane: Structural and rotational dynamics measured with 2D IR and pump–probe experiments. Journal of the American Chemical Society. 2017;**139**(32):11222-11232

[19] Shin JY, Yamada SA, Fayer MD. Influence of water on carbon dioxide and room temperature ionic liquid dynamics: Supported ionic liquid membrane vs the bulk liquid. The Journal of Physical Chemistry. B. 2018;**122**(8):2389-2395

[20] Giammanco CH, Kramer PL, Fayer MD. Ionic liquid versus Li+ aqueous solutions: Water dynamics near bistriflimide anions. The Journal of Physical Chemistry. B. 2016;**120**(37): 9997-10009

[21] Giammanco CH, Yamada SA, Kramer PL, Tamimi A, Fayer MD. Structural and rotational dynamics of carbon dioxide in 1-Alkyl-3-methylimidazolium Bis (trifluoromethylsulfonyl)imide ionic liquids: The effect of chain length. The Journal of Physical Chemistry. B. 2016; **120**(27):6698-6711

[22] Giammanco CH, Kramer PL, Yamada SA, Nishida J, Tamimi A, Fayer MD. Carbon dioxide in an ionic liquid: Structural and rotational dynamics. The Journal of Chemical Physics. 2016;**144**(10):104506

[23] Ren Z, Ivanova AS, Couchot-Vore D, Garrett-Roe S. Ultrafast structure and dynamics in ionic liquids: 2D-IR spectroscopy probes the molecular origin of viscosity. Journal of Physical Chemistry Letters. 2014;**5**(9):1541-1546

[24] Ren Z, Brinzer T, Dutta S, Garrett-Roe S. Thiocyanate as a local probe of ultrafast structure and dynamics in imidazolium-based ionic liquids: Water-induced heterogeneity and cation-induced ion pairing. The Journal of Physical Chemistry. B. 2015;**119**(13): 4699-4712

[25] Terranova ZL, Corcelli SA. Molecular dynamics investigation of the vibrational spectroscopy of isolated water in an ionic liquid. The Journal of Physical Chemistry. B. 2014;**118**(28): 8264-8272

[26] Berquist EJ, Daly CA, Brinzer T, Bullard KK, Campbell ZM, Corcelli SA, et al. Modeling carbon dioxide vibrational frequencies in ionic liquids: I. Ab initio calculations. The Journal of Physical Chemistry. B. 2017;**121**(1): 208-220

[27] Daly CA, Berquist EJ, Brinzer T, Garrett-Roe S, Lambrecht DS, Corcelli SA. Modeling carbon dioxide vibrational frequencies in ionic liquids: II. Spectroscopic map. The Journal of Physical Chemistry. B. 2016;**120**(49): 12633-12642

[28] Zheng J, Kwak K, Fayer MD. Ultrafast 2D IR vibrational echo spectroscopy. Accounts of Chemical Research. 2007;**40**(1):75-83

[29] Park S, Kwak K, Fayer MD. Ultrafast 2D-IR vibrational echo spectroscopy: A probe of molecular dynamics. Laser Physics Letters. 2007;**4**(10):704

[30] Kim H, Cho M. Infrared probes for studying the structure and dynamics of biomolecules. Chemical Reviews. 2013; **113**(8):5817-5847

[31] Fecko CJ, Eaves JD, Loparo JJ, Tokmakoff A, Geissler PL. Ultrafast hydrogen-bond dynamics in the infrared spectroscopy of water. Science. 2003; **301**(5640):1698-1702

[32] Asbury JB, Steinel T, Fayer MD. Hydrogen bond networks: Structure and evolution after hydrogen bond breaking. The Journal of Physical Chemistry. B. 2004;**108**(21):6544-6554

[33] Asbury JB, Steinel T, Stromberg C, Corcelli SA, Lawrence CP, Skinner JL, et al. Water dynamics: Vibrational echo correlation spectroscopy and comparison to molecular dynamics simulations. The Journal of Physical Chemistry. A. 2004; **108**(7):1107-1119

[34] Giammanco CH, Kramer PL, Wong DB, Fayer MD. Water dynamics in 1-Alkyl-3- methylimidazolium tetrafluoroborate ionic liquids. The Journal of Physical Chemistry. B. 2016; **120**(44):11523-11538

[35] Zheng J, Kwak K, Xie J, Fayer MD. Ultrafast carbon-carbon single-bond rotational isomerization in room-temperature solution. Science. 2006; **313**(5795):1951-1955

[36] Biswas A, Mallik BS. Conformational dynamics of aqueous hydrogen peroxide from first principles molecular dynamics simulations. Physical Chemistry Chemical Physics. 2020;**22**(48): 28286-28296

[37] Khalil M, Demirdöven N, Tokmakoff A. Vibrational coherence transfer characterized with Fourier-transform 2D IR spectroscopy. The Journal of Chemical Physics. 2004; **121**(1):362-373

[38] Zanni MT, Hochstrasser RM. Two-dimensional infrared spectroscopy: A promising new method for the time resolution of structures. Current Opinion in Structural Biology. 2001;**11**(5):516-522

[39] Fayer MD. Fast protein dynamics probed with infrared vibrational echo experiments. Annual Review of Physical Chemistry. 2001;**52**:315-356

[40] Merchant KA, Noid WG, Thompson DE, Akiyama R, Loring RF, Fayer MD. Structural assignments and dynamics of the a substates of MbCO: Spectrally resolved vibrational echo experiments and molecular dynamics simulations. The Journal of Physical Chemistry B. 2003;**107**(1):4-7

[41] Fang C, Bauman JD, Das K, Remorino A, Arnold E, Hochstrasser RM. Two- dimensional infrared spectra reveal relaxation of the nonnucleoside inhibitor TMC278 complexed with HIV-1 reverse transcriptase. Proceedings of the National Academy of Sciences of the United States of America. 2008;**105**(5): 1472-1477

[42] Williams IM, Qasim LN, Tran L, Scott A, Riley K, Dutta S. C-D vibration at C2 position of imidazolium cation as a probe of the ionic liquid microenvironment. The Journal of Physical Chemistry. A. 2019;**123**(29): 6342-6349

[43] Kramer PL, Nishida J, Fayer MD. Separation of experimental 2D IR frequency-frequency correlation functions into structural and reorientation-induced contributions. The Journal of Chemical Physics. 2015; **143**(12):124505

[44] Kwak K, Park S, Finkelstein IJ, Fayer MD. Frequency-frequency correlation functions and apodization in two-dimensional infrared vibrational echo spectroscopy: A new approach. The

Journal of Chemical Physics. 2007; **127**(12):124503

[45] Thomaz JE, Lawler CM, Fayer MD. The influence of water on the alkyl region structure in variable chain length imidazolium-based ionic liquid/water mixtures. The Journal of Physical Chemistry. B. 2016;**120**(39):10350-10357

[46] Deàk JC, Rhea ST, Iwaki LK, Dlott DD. Vibrational energy relaxation and spectral diffusion in water and deuterated water. The Journal of Physical Chemistry. A. 2000;**104**(21): 4866-4875

[47] Corcelli SA, Lawrence CP, Asbury JB, Steinel T, Fayer MD, Skinner JL. Spectral diffusion in a fluctuating charge model of water. The Journal of Chemical Physics. 2004; **121**(18):8897-8900

[48] Mallik BS, Semparithi A, Chandra A. Vibrational spectral diffusion and hydrogen bond dynamics in heavy water from first principles. The Journal of Physical Chemistry. A. 2008;**112**(23): 5104-5112

[49] Biswas A, Priyadarsini A, Mallik BS. Dynamics and spectral response of water molecules around tetramethylammonium cation. The Journal of Physical Chemistry. B. 2019;**123**(41):8753-8766

[50] Hutter J, Iannuzzi M, Schiffmann F, VandeVondele J. CP2K: Atomistic simulations of condensed matter systems. Wiley Interdisciplinary Reviews Computational Molecular Science. 2014;**4**(1):15-25

[51] Car R, Parrinello M. Unified approach for molecular dynamics and density- functional theory. Physical Review Letters. 1985;**55**(22):2471-2474

[52] Daubechies I. The wavelet transform, time-frequency localization and signal analysis. IEEE Transactions on Information Theory. 1990;**36**(5): 961-1005

[53] Vela-Arevalo LV, Wiggins S. Time-frequency analysis of classical trajectories of polyatomic molecules. International Journal of Bifurcation and Chaos. 2001;**11**(5):1359-1380

[54] Choudhuri JR, Yadav VK, Karmakar A, Mallik BS, Chandra A. A first-principles theoretical study of hydrogen-bond dynamics and vibrational spectral diffusion in aqueous ionic solution: Water in the hydration shell of a fluoride ion. Pure and Applied Chemistry. 2012;**85**(1):27-40

[55] Mallik BS, Semparithi A, Chandra A. A first principles theoretical study of vibrational spectral diffusion and hydrogen bond dynamics in aqueous ionic solutions: D_2O in hydration shells of $Cl-$ ions. The Journal of Chemical Physics. 2008;**129**(19):194512

[56] Karmakar A, Choudhuri JR, Yadav VK, Mallik BS, Chandra A. A first principles simulation study of vibrational spectral diffusion in aqueous NaBr solutions: Dynamics of water in ion hydration shells. Chemical Physics. 2013; **412**:13-21

[57] Mallik BS, Chandra A. Vibrational spectral diffusion in supercritical D_2O from first principles: An interplay between the dynamics of hydrogen bonds, dangling OD groups, and inertial rotation. The Journal of Physical Chemistry. A. 2008;**112**(51):13518-13527

[58] Mallik BS, Chandra A. An ab initio molecular dynamics study of supercritical aqueous ionic solutions: Hydrogen bonding, rotational dynamics and vibrational spectral diffusion. Chemical Physics. 2011;**387**(1):48-55

[59] Biswas S, Mallik BS. Aqueous solvation of an amide molecule from first principles molecular simulations: Structure, hydrogen bond dynamics and spectral signature. Journal of Molecular Liquids. 2015;**212**:941-946

[60] Biswas S, Mallik BS. Ultrafast vibrational spectroscopy of aqueous solution of methylamine from first principles MD simulations. Chemistry Select. 2017;**2**(1):74-83

[61] Biswas S, Mallik BS. Time-dependent vibrational spectral analysis of first principles trajectory of methylamine with wavelet transform. Physical Chemistry Chemical Physics. 2017;**19**(15):9912-9922

[62] Biswas S, Mallik BS. Aqueous hydroxyl group as the vibrational probe to access the hydrophobicity of amide derivatives. Journal of Molecular Liquids. 2019;**301**:112395

[63] Biswas A, Mallik BS. Structure and stretching dynamics of water molecules around an amphiphilic amide from FPMD simulations: A case study of N,N-dimethylformamide. Journal of Molecular Liquids. 2020;**302**:112524

[64] Biswas S, Mallik BS. Probing the vibrational dynamics of amide bands of N- methylformamide, N, N-dimethylacetamide, and N-methylacetamide in water. Computational and Theoretical Chemistry. 2020;**1190**:113001

[65] Biswas S, Mallik BS. Solvent-mediated dynamics and stretching profile of amide modes: QM/MM simulations of N-methylacetamide in ionic and various molecular liquids. Journal of Molecular Liquids. 2020;**317**:114202

[66] Biswas S, Mallik BS. Vibration spectral dynamics of weakly coordinating water molecules near an anion: FPMD simulations of an aqueous solution of tetrafluoroborate. The Journal of Physical Chemistry. B. 2019; **123**(9):2135-2146

[67] Biswas S, Mallik BS. Heterogeneous occupancy and vibrational dynamics of spatially patterned water molecules. The Journal of Physical Chemistry. B. 2019; **123**(19):4278-4290

[68] Biswas S, Mallik BS. Negligible effect on the structure and vibrational spectral dynamics of water molecules near hydrophobic solutes. Chemistry Select. 2020;**5**(37):11549-11559

[69] Biswas A, Mallik BS. Distinctive behavior and two-dimensional vibrational dynamics of water molecules inside glycine solvation shell. RSC Advances. 2020;**10**(11):6658-6670

[70] Mallik BS, Chandra A. Hydrogen bond dynamics and vibrational spectral diffusion in aqueous solution of acetone: A first principles molecular dynamics study#. Journal of Chemical Sciences. 2012;**124**(1):215-221

[71] Biswas A, Mallik BS. Conformation-induced vibrational spectral dynamics of hydrogen peroxide and vicinal water molecules. Physical Chemistry Chemical Physics. 2021;**23**(11):6665-6676

[72] Kuo IFW, Tobias DJ. Thermal fluctuations of the unusually symmetric and stable superoxide tetrahydrate complex: An *ab initio* molecular dynamics study. The Journal of Physical Chemistry. A. 2002;**106**(46): 10969-10976

[73] Lee HS, Tuckerman ME. Dynamical properties of liquid water from *ab initio* molecular dynamics performed in the complete basis set limit. The Journal of Chemical Physics. 2007;**126**(16):164501

[74] Brézin E, Zinn-Justin J. Finite size effects in phase transitions. Nuclear Physics B. 1985;**257**:867-893

[75] Iglesias Ò, Ritort F, Labarta A. Finite size effects in small particle systems. In: Hadjipanayis GC, editor. Magnetic Storage Systems beyond 2000 [Internet]. Dordrecht: Springer Netherlands; 2001. pp. 363-367. DOI: 10.1007/978-94-010-0624-8_28. (NATO Science Series)

[76] Salacuse JJ, Egelstaff PA. Finite-size effects in molecular dynamics simulations: Intermediate scattering function and velocity of sound. III. Theory and application to a model krypton fluid. Physical Review. E, Statistical, Nonlinear, and Soft Matter Physics. 2001;**64**(5 Pt 1):051201

[77] Godec A, Metzler R. Finite-time effects and ultraweak ergodicity breaking in superdiffusive dynamics. Physical Review Letters. 2013;**110**(2): 020603

[78] Yang G, Nilsson F, Schubert DW. A study of finite size effects and periodic boundary conditions to simulations of a novel theoretical self-consistent mean-field approach. Macromolecular Theory and Simulations. 2019;**28**(5):1900023

[79] Corcelli SA, Lawrence CP, Skinner JL. Combined electronic structure/molecular dynamics approach for ultrafast infrared spectroscopy of dilute HOD in liquid H_2O and D_2O. The Journal of Chemical Physics. 2004; **120**(17):8107-8117

[80] Auer B, Kumar R, Schmidt JR, Skinner JL. Hydrogen bonding and Raman, IR, and 2D-IR spectroscopy of dilute HOD in liquid D_2O. PNAS. 2007; **104**(36):14215-14220

[81] Biswas A, Mallik BS. Ultrafast aqueous dynamics in concentrated electrolytic solutions of lithium salt and ionic liquid. The Journal of Physical Chemistry. B. 2020;**124**(44):9898-9912

[82] Dinç E, Yazan Z. Wavelet transform-based UV spectroscopy for pharmaceutical analysis. Frontiers in Chemistry. 2018;**6**:503

[83] Perrier V, Philipovitch T, Basdevant C. Wavelet spectra compared to Fourier spectra. Journal of Mathematical Physics. 1995;**36**(3): 1506-1519

[84] Drozdov A, Pomortsev I, Tyutyukin K, Baloshin Y. Comparison of wavelet transform and Fourier transform applied to analysis of non-stationary processes. Наносистемы: физика, химия, математика. 2014;**5**(3): 363-373

[85] Laskar J, Froeschlé C, Celletti A. The measure of chaos by the numerical analysis of the fundamental frequencies. Application to the standard mapping. Physica D: Nonlinear Phenomena. 1992; **56**(2):253-269

[86] Laskar J. Frequency analysis for multi-dimensional systems. Global dynamics and diffusion. Physica D: Nonlinear Phenomena. 1993;**67**(1): 257-281

[87] Shatkay H. The Fourier Transform - a Primer. Providence, RI, USA: Brown University; 1995

[88] Carmona R, Hwang WL, Torresani B. Practical Time-Frequency Analysis: Gabor and Wavelet Transforms, with an Implementation in S. USA: Academic Press; 1998. 494 p

[89] Bankura A, Karmakar A, Carnevale V, Chandra A, Klein ML. Structure, dynamics, and spectral diffusion of water from first-principles

molecular dynamics. Journal of Physical Chemistry C. 2014;**118**(50):29401-29411

[90] Yadav VK, Chandra A. First-principles simulation study of vibrational spectral diffusion and hydrogen bond fluctuations in aqueous solution of N-methylacetamide. The Journal of Physical Chemistry. B. 2015; **119**(30):9858-9867

[91] MacFarlane DR, Forsyth M, Howlett PC, Pringle JM, Sun J, Annat G, et al. Ionic liquids in electrochemical devices and processes: Managing interfacial electrochemistry. Accounts of Chemical Research. 2007;**40**(11): 1165-1173

[92] Wishart JF. Energy applications of ionic liquids. Energy & Environmental Science. 2009;**2**(9):956-961

[93] Condemarin R, Scovazzo P. Gas permeabilities, solubilities, diffusivities, and diffusivity correlations for ammonium-based room temperature ionic liquids with comparison to imidazolium and phosphonium RTIL data. Chemical Engineering Journal. 2009;**147**(1):51-57

[94] Seifert E. OriginPro 9.1: Scientific data analysis and graphing software—Software review. Journal of Chemical Information and Modeling. 2014;**54**(5): 1552-1552

[95] Fulfer KD, Kuroda DG. Solvation structure and dynamics of the lithium ion in organic carbonate-based electrolytes: A time-dependent infrared spectroscopy study. Journal of Physical Chemistry C. 2016;**120**(42):24011-24022

[96] Kramer PL, Giammanco CH, Fayer MD. Dynamics of water, methanol, and ethanol in a room temperature ionic liquid. The Journal of Chemical Physics. 2015;**142**(21):212408

[97] Stangret J, Gampe T. Ionic hydration behavior derived from infrared spectra in HDO. The Journal of Physical Chemistry. A. 2002;**106**(21):5393-5402

[98] Wong DB, Giammanco CH, Fenn EE, Fayer MD. Dynamics of isolated water molecules in a sea of ions in a room temperature ionic liquid. The Journal of Physical Chemistry. B. 2013; **117**(2):623-635

[99] Fayer MD, Moilanen DE, Wong D, Rosenfeld DE, Fenn EE, Park S. Water dynamics in salt solutions studied with ultrafast 2D IR vibrational echo spectroscopy. Accounts of Chemical Research. 2009;**42**(9):1210-1219

[100] Giammanco CH, Kramer PL, Fayer MD. Dynamics of dihydrogen bonding in aqueous solutions of sodium borohydride. The Journal of Physical Chemistry. B. 2015;**119**(8):3546-3559

[101] Giammanco CH, Wong DB, Fayer MD. Water dynamics in divalent and monovalent concentrated salt solutions. The Journal of Physical Chemistry B. 2012;**116**(46):13781-13792

[102] Priyadarshini A, Biswas A, Chakraborty D, Mallik BS. Structural and thermophysical anomalies of liquid water: A tale of molecules in the instantaneous low- and high-density regions. The Journal of Physical Chemistry. B. 2020;**124**(6):1071-1081

[103] Moilanen DE, Wong D, Rosenfeld DE, Fenn EE, Fayer MD. Ion–water hydrogen- bond switching observed with 2D IR vibrational echo chemical exchange spectroscopy. PNAS. 2009;**106**(2):375-380

[104] Ji M, Odelius M, Gaffney KJ. Large angular jump mechanism observed for hydrogen bond exchange in aqueous

perchlorate solution. Science. 2010; **328**(5981):1003-1005

[105] Maginn EJ. Molecular simulation of ionic liquids: Current status and future opportunities. Journal of Physics. Condensed Matter. 2009;**21**(37):373101

[106] Floisand DJ, Miller TC, Corcelli SA. Dynamics and vibrational spectroscopy of alcohols in ionic liquids: Methanol and ethanol. The Journal of Physical Chemistry. B. 2019;**123**(38):8113-8122

[107] Kananenka AA, Yao K, Corcelli SA, Skinner JL. Machine learning for vibrational spectroscopic maps. Journal of Chemical Theory and Computation. 2019;**15**(12):6850-6858

[108] Kwac K, Cho M. Machine learning approach for describing vibrational solvatochromism. The Journal of Chemical Physics. 2020;**152**(17):174101

[109] Schröder C. Comparing reduced partial charge models with polarizable simulations of ionic liquids. Physical Chemistry Chemical Physics. 2012; **14**(9):3089-3102

[110] Ishizuka R, Matubayasi N. Self-consistent determination of atomic charges of ionic liquid through a combination of molecular dynamics simulation and density functional theory. Journal of Chemical Theory and Computation. 2016;**12**(2):804-811

[111] Biswas A, Mallik BS. Microheterogeneity-induced vibrational spectral dynamics of aqueous 1-Alkyl-3-methylimidazolium tetrafluoroborate ionic liquids of different cationic chain lengths. The Journal of Physical Chemistry. B. 2022;**126**(29):5523-5533

[112] Biswas A, Mallik BS. Vibrational spectral dynamics and ion-probe interactions of the hydrogen-bonded liquids in 1-ethyl-3-methylimidazolium bis(trifluoromethylsulfonyl)imide. Chemical Physics. 2022;**559**:111519

[113] Biswas A, Mallik BS. Multiple ensembles of the hydrogen-bonded network in ethylammonium nitrate versus water from vibrational spectral dynamics of SCN−probe. ChemPhysChem. 2022;**23**(23): e202200497

[114] Biswas A, Mallik BS. Molecular simulation-guided spectroscopy of imidazolium-based ionic liquids and effects of methylation on ion-cage and -pair dynamics. The Journal of Physical Chemistry. B. 2022;**126**(43):8838-8850

[115] Biswas A, Mallik BS. Direct correlation between short-range vibrational spectral diffusion and localized ion-cage dynamics of water-in-salt electrolytes. The Journal of Physical Chemistry. B. 2023;**127**(1):236-248

[116] Liang C, Kwak K, Cho M. Revealing the solvation structure and dynamics of carbonate electrolytes in lithium-ion batteries by two-dimensional infrared spectrum modeling. Journal of Physical Chemistry Letters. 2017;**8**(23):5779-5784

[117] Lim J, Park K, Lee H, Kim J, Kwak K, Cho M. Nanometric water channels in water-in-salt lithium ion battery electrolyte. Journal of the American Chemical Society. 2018; **140**(46):15661-15667

[118] Baldelli S, Schnitzer C, Simonelli D. Aqueous solution/air interfaces probed with sum frequency generation spectroscopy. The Journal of Physical Chemistry. B. 2002;**106**(21):5313-5324

[119] Paul S, Chandra A. Liquid-vapor interfacial properties of water-ammonia mixtures: Dependence on ammonia

concentration. The Journal of Chemical Physics. 2005;**123**(17):174712

[120] Paul S, Chandra A. Hydrogen bond properties and dynamics of liquid−vapor interfaces of aqueous methanol solutions. Journal of Chemical Theory and Computation. 2005;**1**(6): 1221-1231

[121] Yuan R, Yan C, Nishida J, Fayer MD. Dynamics in a water interfacial boundary layer investigated with IR polarization-selective pump–probe experiments. The Journal of Physical Chemistry. B. 2017;**121**(17): 4530-4537

[122] Kundu A, Verma PK, Cho M. Water structure and dynamics in the stern layer of micelles: Femtosecond mid-infrared pump-probe spectroscopy study. The Journal of Physical Chemistry. B. 2019; **123**(25):5238-5245

[123] Díaz-Visurraga J, Daza C, Pozo C, Becerra A, von Plessing C, García A. Study on antibacterial alginate-stabilized copper nanoparticles by FT-IR and 2D-IR correlation spectroscopy. International Journal of Nanomedicine. 2012;7:3597-3612

[124] López-Lorente ÁI, Mizaikoff B. Recent advances on the characterization of nanoparticles using infrared spectroscopy. TrAC Trends in Analytical Chemistry. 2016;**84**:97-106

[125] Ghosh A, Prasad AK, Chuntonov L. Two-dimensional infrared spectroscopy reveals molecular self-assembly on the surface of silver nanoparticles. Journal of Physical Chemistry Letters. 2019;**10**(10): 2481-2486

[126] Scheurer C, Piryatinski A, Mukamel S. Signatures of β-peptide unfolding in two-dimensional vibrational echo spectroscopy: A simulation study. Journal of the American Chemical Society. 2001; **123**(13):3114-3124

[127] Mu Y, Stock G. Conformational dynamics of trialanine in water: A molecular dynamics study. The Journal of Physical Chemistry. B. 2002;**106**(20): 5294-5301

[128] Hume S, Hithell G, Greetham GM, Donaldson PM, Towrie M, Parker AW, et al. Measuring proteins in H_2O with 2D-IR spectroscopy. Chemical Science. 2019;**10**(26):6448-6456

[129] Ramos S, Horness RE, Collins JA, Haak D, Thielges MC. Site-specific 2D IR spectroscopy: A general approach for the characterization of protein dynamics with high spatial and temporal resolution. Physical Chemistry Chemical Physics. 2019;**21**(2):780-788

[130] Stenger J, Madsen D, Dreyer J, Nibbering ETJ, Hamm P, Elsaesser T. Coherent response of hydrogen bonds in liquids probed by ultrafast vibrational spectroscopy. The Journal of Physical Chemistry. A. 2001;**105**(13):2929-2932

[131] Asbury JB, Steinel T, Stromberg C, Gaffney KJ, Piletic IR, Goun A, et al. Hydrogen bond dynamics probed with ultrafast infrared heterodyne-detected multidimensional vibrational stimulated echoes. Physical Review Letters. 2003; **91**(23):237402

[132] Kübel J, Lee G, Ooi SA, Westenhoff S, Han H, Cho M, et al. Ultrafast chemical exchange dynamics of hydrogen bonds observed via Isonitrile infrared sensors: Implications for biomolecular studies. Journal of Physical Chemistry Letters. 2019;**10**(24): 7878-7883

Chapter 2

Refractive Index in Ionic Liquids: Beyond the D Line

Yago Arosa, Carlos Damián Rodríguez-Fernández, Elena López Lago and Raúl De la Fuente

Abstract

In the last decade, the intrinsic tunability of ionic liquids has attracted interest well beyond the disciplines that inaugurated the research on these materials at the beginning of the 1990s. One of these emerging fields is photonics, where the possibility of designing materials with task-specific optical responses is promising for a large number of potential applications. Among the optical properties of ionic liquids, refractive index has been the subject of a thorough investigation because of its relevance in the performance of most optical devices. In this chapter, we address the recent experimental and computational advances on the characterization of the refractive index of ionic liquids and the relations of this magnitude with their structure. Furthermore, we provide an important review of works introducing ionic liquids as essential optical materials in diverse applications.

Keywords: ionic liquids, photonics, optical properties, refractive index, chromatic dispersion, electronic polarizability

1. Introduction

It is well known that ionic liquids (ILs) are very stable ionic materials with a large intrinsic tunability. From the very beginning of the research on ILs, their suitability to be used as designer solvents (or smart materials) was recognized, i.e., the possibility of synthetizing ILs with tailored physicochemical properties to fulfill task-specific requirements. The idea of easily obtaining solvents with specific responses made ILs the subject of an early intensive research in fields as diverse as electrochemistry, catalysis or separation science. Indeed, over the last few years, the tunability inherent to ILs has attracted further attention in new fields, such as photonics.

Designing stable (and non-evaporating liquids) with tailored optical responses is promising for different photonic applications. Among the wide variety of existing optical properties, refractive index is the most fundamental magnitude in any material. In short, it governs the phase velocity of light in a medium and, together with its dispersion; it determines the possible applications of such a material as an optical element. The pioneering measurements of refractive index in ILs were performed as a part of their necessary physicochemical characterization. However, since the possibility of effectively tuning the refractive index in ILs was recognized, the number of works dealing with this magnitude from an optical perspective has been continuously growing.

In this chapter, the recent research on the refractive index of ILs is reviewed, including the most outstanding experimental and computational milestones in the field. In addition, the basic tools for tuning the refractive index are summarized and some examples of practical implementation of ILs in optical devices are addressed. The short-names used for the ILs mentioned in the text can be consulted in the "Abbreviations" section at the end of this chapter.

2. Experimental determination of the refractive index

2.1 Single-wavelength measurements

A large part of the refractive index measurements in ILs are performed at a single wavelength, the sodium D-line, λ_D = 589.6 nm, as a part of the routinary characterization of the physicochemical properties of new ion combinations or functionalized species. Starting with the pioneering studies by Bonhôte et al. and Huddleston et al. [1, 2], numerous articles were published dealing with the characterization of this magnitude and its dependence on temperature (typically between 293 and 333 K). A representative summary of the research up to 2014 can be found in Ref. [3]. The published data clearly demonstrates the ability to tune the refractive index as a function of the selected counterions, by mixing ILs or dissolving them in other miscible liquids. ILs with refractive indices between 1.60 and 2.08 RIU, comparable to those of optical glasses, were already synthesized in that moment. However, many of them were not transparent in the visible range, which limited their application in imaging or guiding systems.

A few years later, Kayama and co-workers [4] synthesized colorless and transparent ILs with refractive indices greater than 1.60 RIU by combining $[BzC_1im]^+$, $[P_{ppp6}]^+$, and $[C_4C_1imI_2]^+$ cations with $[Sac]^-$, $[Bz\text{-}I\text{-}SO_3]^-$, $[Br]^-$, and $[N(CN)_2]^-$ anions. The strategy followed for the design of these high index compounds was based on the combination of ions with high molar refractions. In this case, they chose to use cations with aromatic structures and to incorporate iodine in the anions.

Other researchers explored the effect on the refractive index of previously unexplored functional groups inserted into the cation. For example, relatively high refractive indices were obtained by Gonfa et al. in 2015 [5] on the next ILs based on the thiocyanate anion combined with: $[C_4C_1im]^+$, $[C_2CNC_4im]^+$, $[C_2CNAim]^+$, $[C_2CNBzim]^+$, and $[C_2CNHeim]^+$. The refractive indices of this set of ILs decreases according to the series: $[C_2CNHeim][SCN]$ > $[C_2CNBzim][SCN]$ > $[C_2CNAim][SCN]$ > $[C_2CNC_4im][SCN]$ > $[C_4C_1im][SCN]$, being their respective values 1.5682, 1.5666, 1.5663, 1.5483, and 1.5391 RIU at 298 K. In this regard, Ullah et al. [6] characterized the IL $[C_2CNC_4im][N(CN)_2]$, which is composed by one of the previous cations paired with a different anion (dicyanamide), in the 293–353 K temperature range. The refractive index values at those temperatures varied from 1.5195 to 1.5061 RIU, being significantly lower than those obtained by replacing the anion by $[SCN]^-$.

Bhattacharjee et al. [7] performed a study in a set of ILs based on the n-alkylphosphonium cations: $[P_{i(444)1}][Tos]$, $[P_{4441}][C_1SO_4]$, $[P_{4442}][(C_2H_5O)_2PO_2]$, and $[P_{8888}][Br]$. The trend observed in the refractive indices is as follows: $[P_{i(444)1}][Tos]$ > $[P_{8888}][Br]$ > $[P_{4441}][C_1SO_4]$ > $[P_{4442}]$ $[(C_2H_5O)_2PO_2]$, being their values 1.51233, 1.46912, 1.45983, and 1.48262 RIU at 318 K. Rodil et al. [8] characterized the refractive index of other phosphonium-based ILs in the temperature range from 283 to 343 K, namely $[P_{4441}][Cl]$, $[P_{4441}][NTf_2]$, $[P_{4441}][C_1SO_4]$, but also $[C_1py][C_1SO_4]$, and $[C_4C_1im][C_1SO_4]$. They observed that the refractive index increased according to the following sequence: $[P_{4441}][NTf_2] < [P_{4441}][C_1SO_4] < [C_4C_1im][C_1SO_4] < [P_{4441}][Cl] < [C_1py][C_1SO_4]$. The values obtained at 293 K are, respectively: 1.4374, 1.4755, 1.4804, 1.5037, and 1.5139 RIU. Hence, cations based on imidazolium or pyridinium species tend to produce ILs with higher refractive index than their phosphonium-based equivalents. On the other hand, Othman et al. [9] focused on ILs based on ammonium cations instead of phosphonium ones. Particularly, they combined diethylammonium and dibutylammonium with alkanoate anions of variable length, $[N_{22}][C_5CO_2]$, $[N_{22}][C_6CO_2]$, $[N_{22}][C_7CO_2]$, $[N_{44}][C_5CO_2]$, $[N_{44}][C_6CO_2]$, and $[N_{44}][C_7CO_2]$. The refractive indices were measured at temperatures ranging from 293 to 333 K. The importance of these liquids lies in their promising ability to absorb CO_2 under typical laboratory conditions. They observed that at 298 K the refractive index varied between 1.43069 and 1.44354 RIU, increasing with the alkyl chain length of the cation and anion.

Lee et al. [10] focused on the so-called Good's buffer ILs (GB-ILs), of potential use in biotechnology thanks to their self-buffering capacity and biocompatibility. Specifically, they studied all the combinations between $[N_{4444}]^+$, $[P_{4444}]^+$, and $[Ch]^+$ with $[MOPSO]^-$, $[BES]^-$, $[TAPSO]^-$, and $[CAPSO]^-$. The obtained refractive indices grow according to the following cationic order: $[N_{4444}]^+ < [P_{4444}]^+ < [Ch]^+$, and the following anionic order: $[MOPSO]^- < [CAPSO]^- < [BES]^- < [TAPSO]^-$. The measured values for the refractive index of these GB-ILs lie in a narrow window (1.47–1.52 RIU) over the entire temperature range employed (288–353 K). On the other hand, Marcinkowski et al. [11] reported refractive index values for ILs based on amino acids, namely they focused in the N-acetyl-L-amino acid ions $[Ala]^-$, $[Val]^-$, $[Leu]^-$, and $[Ile]^-$ combined with the morpholinium-based cations $[Mor_{1,k}]^+$ of variable alkyl chain (k = 2, 3, 4, 6, 8, where k is the number of carbons in the chain). These ILs show low refractive indices, between 1.481 and 1.488 RIU at 318 K.

Khan et al. [12] studied the refractive index of dicationic ILs (DILs) as a further tuning variable. They focused on four ILs based on the $[C_4(C_1im)_2]^{2+}$ cation combined with the anions $[HSO_4]^-$, $[C_1SO_3]^-$, $[OTf]^-$, and $[\rho\text{-TSA}]^-$. Refractive index values were measured in the temperature range from 293 to 323 K. The decreasing order of refractive indices of these DILs is: $[C_4(C_1im)_2][OTf]_2 > [C_4(C_1im)_2][HSO_4]_2 > [C_4(C_1im)_2][C_1SO_4]_2 > [C_4(C_1im)_2][\rho\text{-TSA}]_2$. The lowest value is shown by $[C_4(C_1im)_2][OTf]_2$ while the highest value is that of $[C_4(C_1im)_2][\rho\text{-TSA}]_2$, whose anion exhibits larger aromaticity and bulkier structure than the rest. In addition, and according to literature, the values of the refractive indices for these DILs are higher than their monocationic equivalents. For instance, the values of the refractive indices of $[C_4C_1im][C_1SO_3]$ and $[C_4C_1im][OTf]$ are, respectively, 1.4792 and 1.4368 RIU, which are lower than those of $[C_4(C_1im)_2][C1SO_3]_2$ and $[C_4(C_1im)_2][OTf]_2$.

As depicted by the previous works, a large part of the research on refractive index has been done in neat ILs. However, some authors focused on mixtures of ILs with other solvents. A relevant contribution is that of Branco et al. [13], who determined the refractive index of index of four ILs: $[C_6C_1im][NTf_2]$, $[C_{10}C_1im][NTf_2]$, $[C_4C_1im][N(CN)_2]$, and $[P_{666,14}][N(CN_2)]$: mixed with Poly(ethylene glycol) 200 (PEG 200) at 298 K. They obtained refractive index values between 1.5096 and 1.4311 RIU. These measurements were combined with existing volumetric data to obtain molar refraction at different mole fractions. They showed that the molar refraction is highly dependent on molar volume although it follows a linear trend upon mixing. Oppositely, they also show that the refractive index follows a nonlinear trend with the molar composition of the mixture. Other important work is that of Wang et al. [14], who studied the refractive index of an asymmetrical DIL, $[piC_5py][NTf_2]_2$, and its binary mixtures with acetonitrile at temperatures from 293 to 323 K. In this case, the achieved refractive index range goes from 1.3297 to 1.4514 RIU.

2.2 Wavelength-dependent measurements

Although most refractive index measurements in ILs concern λ_D, it is increasingly common to find articles aimed to determine this magnitude over broader bands in the visible range and adjacent spectral regions. These wavelength-dependent measurements are spurred by the emergence of different optical applications (see Section 4), in which material chromatic dispersion, the derivative of the refractive index with respect to wavelength, is of crucial importance. Since ILs usually present high absorption below 300 nm, these measurements often concern longer wavelengths in the UV, the whole visible region and the NIR up to about 2 microns.

To our knowledge, the first refractive index measurements of ILs at different wavelengths were performed in 2011 by Calixto and co-workers [15]: they studied six ILs, mainly from the imidazolium family, at six wavelengths in the range of 400–650 nm. The authors indicate that they used a spectroscope to carry out the measurements, but do not give further details about the apparatus. In 2014, Novoa et al. [16] used a multi-wavelength Abbe refractometer (Atago DR-2) to measure the refractive index of a set of five ammonium-based and five 1-butyl-3-methylimidazolium-based ILs at six wavelengths in the 475–680 nm range. Interpolation of these data using a Cauchy dispersion formula allowed them to determine the refractive index at other wavelengths in that range.

In 2017, Chiappe et al. [17] performed systematic studies of wavelength-dependent refractive index in 1-alkyl-3-methylimidazolium-based ILs, with alkyl chains of length k = 1, 4 and 6, combined with three phosphorus-containing anions, $[(C_1)_2PO_2]^-$, $[C_1C_1OPO_2]^-$, and $[C_1OHPO_2]^-$. Measurements were performed at five wavelengths (450, 532, 632.8, 964, and 1551 nm) in the temperature range from 353 to 443 K using prism coupling as the method for the refractive index determination. A linear decrease of refractive index with temperature was observed, meaning that the thermo-optic coefficient, which is the variation of refractive index with temperature, dn/dT, is constant in the measuring range. Regarding dispersion, typical curves of decreasing refractive index with increasing wavelength were obtained (normal dispersion) and the refractive index values within the spectral region of interest were interpolated at T = 393 K using a five-term Cauchy equation:

$$n^2(\lambda, 393\,K) = A_0 + A_1\lambda^2 + \frac{A_2}{\lambda^2} + \frac{A_3}{\lambda^4} + \frac{A_4}{\lambda^6}, \quad (1)$$

while, for temperature, they considered the following dependence:

$$n(\lambda, T) = n(\lambda, 393\,K) + (T - 393\,K)\frac{dn}{dT}. \quad (2)$$

Regarding the structure of the ILs, the authors observed that, for these particular ILs, refractive index decreases with the length of the alkyl chain of the imidazolium cation. Furthermore, they investigated the effect of each anion on refractive index by means of *ab-initio* calculations.

The same group and using the same method [18] measured the dispersion of seven ILs of the levulinate family, also at the same wavelengths and at a similar temperature range. In this case, the interpolation of the dispersion curves was done using the Sellmeier relation:

$$n^2(\lambda, T) = 1 + \frac{S(T)\lambda^2}{\lambda^2 - \lambda_{uv}^2} + d \cdot \lambda^2, \tag{3}$$

with a linear relationship of S on T. They also determined the group index, which is defined as $n_g(\lambda) = n - \lambda \frac{dn}{d\lambda}$. At the microscopic level, the dependence of the electronic polarizability on λ and T was analyzed.

Wu et al. [19] applied ellipsometry to measure the refractive index of six commercial ILs with values around 1.40 RIU ($[C_4C_1im][PF_6]$, $[C_4C_1im][BF_4]$, $[C_4C_1im][NTf_2]$, $[C_4C_1im][OTf]$, $[P_{666,14}][NTf_2]$, and $[P_{666,14}][Cl]$) in addition to two high refractive index ILs (n > 1.70 RIU): $[C_2C_1im][BrI_2]$ and $[C_2C_1im][I_5]$. Employing this technique allowed them to measure the extinction rate associated to absorption. For the low refractive index ILs, they used a two-term Sellmeier relation to describe their material dispersion. They did not use any relationship for the high refractive index ILs, as the dispersion curve was more complex as a consequence of the higher value of the extinction coefficient. Furthermore, these authors demonstrated that the evaluated ILs can be used as immersion liquids in photonic crystals. Also using ellipsometry, Rola et al. [20] measured the refractive index of solid layers of three polymerized ILs, $[AllC_1im][Cl]$, $[AllC_1im][NTf_2]$, and $[AllC_1C_1im][NTf_2]$. These layers were fabricated by electron beam patterning to be used as planar photonic microcomponents. The refractive indices of the neat ILs were also measured, but, in this case, using a spectroscopy-based method. Indeed, what they did was to study the interference patterns produced by the liquids when introduced within a commercial micrometer-thick cuvette. The refractive index curves of both neat and polymerized ILs were fitted using a three-term Cauchy relation in the spectral range from 300 to 1100 nm. The authors observed that polymerization increases the refractive index from values lower than 1.6 RIU to higher values (increments in the range 0.1 to 0.4 RIU).

Spectroscopic techniques were also used by Elmahdy [21] to measure the refractive index of two ILs, $[C_4C_1im][BF_4]$ and $[C_4C_1im][OTf]$. Transmittance (T) and absorbance (A) measurements in the spectral range from 200 to 1100 nm allowed the determination of the refractive index as follows:

$$n = \frac{1+R}{1-R} + \sqrt{\frac{4R}{(1-R)^2} - \kappa^2}, \tag{4}$$

with κ being the extinction coefficient and $R = 1 - \sqrt{Te^A}$. For both ILs, they observed a refractive index maximum caused by an absorption peak located around 300 nm. In order to reproduce the curve, they made different fits for wavelengths below and above this absorption maximum.

De la Fuente group made extensive use of a technique called Refractive Index Spectroscopy by Broadband Interferometry (RISBI), in which the spectral interference between a beam passing through a cell with the sample and a reference beam is resolved. In 2017, they measured the refractive index of 14 ILs belonging to the imidazolium family [22]. The spectral range of measurement was between 400 and 1000 nm and the temperature was kept constant at 300 K. A five-term Cauchy relation, similar to that of Eq. (1), was used to describe the refractive index curves. The absorption of water by the samples was taken into account, considering the solution as an ideal binary mixture in which the refractive index is given by Newton's equation.

In 2018 the studies continued [23], this time considering a temperature range from 298 to 323 K. In this case, the ILs were divided into four different groups according to their common cations or anions. This arrangement allowed them to analyze the dependence of the refractive index on the length of the alkyl chain of the cation. On the other hand, they employed a single-term Sellmeier formula, similar to that of Eq. (3), to fit the refractive index curves. Using this expression, the authors were able to recognize that the variations of the squared refractive index with temperature and wavelength are independent. In addition, the resonance wavelength of the model turned out to be independent of both alkyl chain length and temperature.

Their work continued in 2020, material dispersion was considered for 20 different ILs and the characterization was extended by incorporating the analysis of higher order dispersion and the thermo-optic coefficient dependence on wavelength [24]. Furthermore, in other work [25], the spectral range of measurement was further increased to include the region from 300 to 1500 nm, as shown in **Figure 1**, while still considering different temperatures within the interval of 293–313 K. In order to fit the refractive index curves in such a broad spectral range, it was necessary to use a Sellmeier dispersion relation with three terms. Again, the authors observed that the wavelengths of resonance were independent of temperature and alkyl chain length:

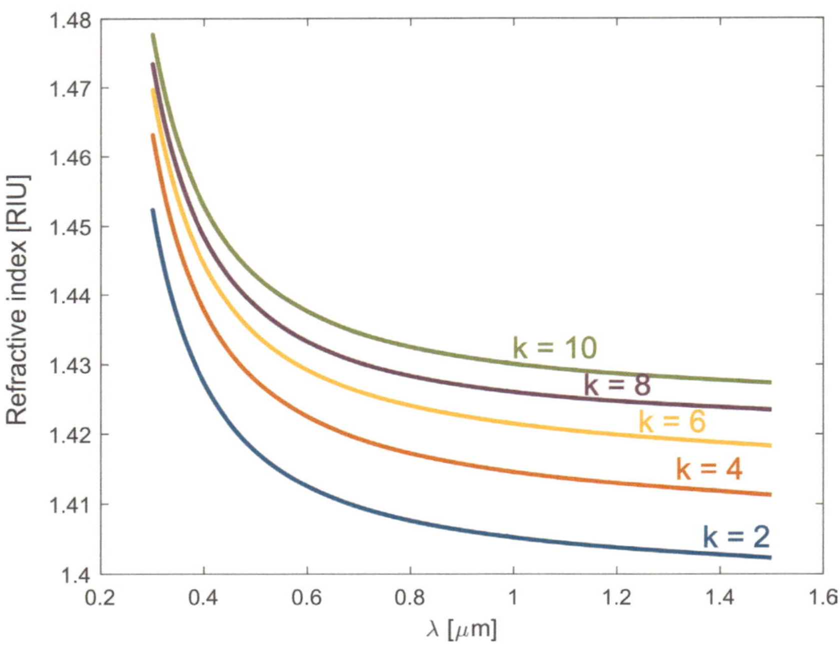

Figure 1.
Refractive index of the [$C_kC_1$im][BF_4] family of ILs for alkyl lengths of k = 2, 4, 6, 8, and 10 in the spectral range of 300–1500 nm at 298 K. Data extracted from Ref. [25].

$$n^2(\lambda, T) = 1 + \left(c_1' + c_1''\Delta T\right)\frac{\lambda^2}{\lambda^2 - \lambda_1^2} + \frac{c_2\lambda^2}{\lambda^2 - \lambda_1^2} + c_3\lambda^2. \tag{5}$$

To analyze the dependence of the refractive index on the alkyl chain length (k), the Lorentz-Lorenz equation was considered:

$$R(\lambda, k) = V(T, k) \cdot \frac{n^2(\lambda, T, k) - 1}{n^2(\lambda, T, k) + 2}, \tag{6}$$

where R is the molar refraction of the IL and V its molar volume. With this expression, the temperature dependence of refractive index was separated from the spectral dependence. This was possible because the thermal contributions arise from the changes of molar volume with temperature (thermal expansion coefficient) while the spectral ones arise from the dependence of electronic polarizability on wavelength. Furthermore, the authors determined that both parameters vary linearly with k.

3. Predictive models

3.1 Prediction based on empirical models

Empirical models are intensively used in the field of ILs to complete the gaps within the available experimental data and to compare disparate compounds. These models, which at times bereft of a theoretical underpinning, venture into the realm of fitting to faithfully represent values through the utilization of elementary equations, such as linear fits or low-order polynomials.

Works dealing with empirical prediction of wavelength-dependent refractive index are limited to Sellmeier formulas and Cauchy relations within the spectral range of measurement. In contrast, the temperature-induced variation of the refractive index finds more extensive documentation within the extant literature. Multiple researchers not only articulate refractive index values along the λ_D at sundry temperatures but also engage in the calculation of the thermo-optic coefficient. This coefficient, generally dependent on temperature for a given material, lends itself to approximation within a confined dynamic range through linear regression, notably applicable to a large number of ILs.

Concurrently, important efforts are done to investigate the correlation between these adjustments and the chemical architecture of ILs. The work by Gardas et al. [26] serves as a compendium, presenting diverse thermo-physical properties modeled through semi-empirical fits, among them, the variation of refractive index with temperature. This group proffers a model to predict the refractive index of ILs which is encapsulated by:

$$n_D = A_{nD} - B_{n_D}T, \tag{7}$$

where the coefficients A_{n_D} and B_{n_D} aggregate the contributions of molecular blocks defined by the authors (cations, anions and carbons in alkyl chains). This analytical pursuit, echoed in other works such as those of Sattari et al. [27] and Ullah et al. [6], exemplifies a kind of strategy based on building the refractive index (and other physicochemical properties) from the contribution of molecular fragments.

Notably, Gardas et al. refrain themselves from an explicit development to discern the impact of each type of fragment in the thermo-optic coefficient. This subsequent elaboration, possibly arising from the theoretical derivation (as alluded to in the text), illuminates the limited influence of polarizability on the thermo-optic coefficient, with the latter predominantly governed by the coefficient of thermal expansion [25]. Despite the foregoing, Gardas' investigation exhibits the occasional sufficiency of an empirical model. Deviations ascertained for a cohort of 24 imidazolium-based ILs reveal relative deviations in refractive index below 0.5% for the 93% of the values.

3.2 Prediction based on electronic polarizability and the Lorenz-Lorenz equation

Semiempirical models are useful ways to rationalize the measurements of refractive index in ILs. However, while the resources to conduct experimental research are limited, the number of possible ion combinations is virtually infinite. For this reason, simulations are especially powerful tools to study the refractive index in these materials. The strategies for computing the refractive index of ILs are diverse, however, they are always based on one of these two approaches: (i) finding the electronic polarizability and using the Lorentz-Lorenz equation to obtain the refractive index or (ii) training an algorithm with experimental refractive indices to predict the refractive index of new species. All the strategies based on electronic polarizability have in common that they exploit the additivity of this magnitude in terms of atoms, molecular groups or ions. However, electronic polarizability can be obtained following very different paths. A large number of works obtain it from *ab initio* calculations based on Density Functional Theory (DFT) or Møller-Plesset Perturbation theory (MP), while others rely on obtaining it from the statistical analysis of experimental data.

Possibly, the work of 2009 of Izgorodina et al. [28] is the first one comparing experimental refractive indices with *ab initio* predictions derived from static electronic polarizabilities (wavelength independent) and molecular volumes. The work itself is not devoted to refractive index but to explore the possible existence of an ionic polarization component in the dielectric constant of ILs. Despite it, they showed that the experimental refractive index at λ_D can be approximately estimated by using the Lorentz-Lorenz equation, Eq. (6), and assuming that the electronic polarizability of the IL is approximately that of the composing ions.

Some years afterwards, in 2013, Bica et al. [29] published the first work completely focused on the estimation of the refractive index by means of computational methods. In this case, the electronic polarizability data was obtained from statistical analysis using the Lorentz-Lorenz equation and experimental refractive indices (at λ_D) and densities. The authors assumed atomic additivity in both electronic polarizability and molar volume, and performed a designed regression analysis to obtain the contributions of individual atoms to these magnitudes. From them, they estimated the electronic polarizability, molar volume and refractive index of a huge number of imidazolium-based ILs. While this model is restricted to ILs containing the imidazolium cation, it is able to provide refractive index with a mean absolute deviation of 0.0035 RIU. Indeed, deviation is lower for most of the ionic combinations, but the overall performance of the model is hindered by its limited ability to reproduce the refractive index of ILs containing the dicyanamide anion.

Another important result of the same work of Bica et al. is that it explains why in some ILs the refractive index rises with the length of alkyl chains while, in others, the trend is the opposite. According to the authors, the key parameter to explain these different behaviors is the Molecular Polarizability Density (MPD) of the -CH_2- unit.

The MPD is nothing more than the ratio between electronic polarizability and volume in a molecular fragment or molecule, $MPD = \alpha_{mol}/V_{mol}$. In accordance with the Lorentz-Lorenz equation, Eq. (6), this ratio is just what determines the value of the refractive index in a liquid. For -CH_2- units, it is 0.0661, in the CGS system (dimensionless) or 7.3546×10^{-12}, C V^{-1} m^{-1} in the SI system. When the MPD of the combination of a cation and an anion is lower than that of the -CH_2- units, increasing the length of an alkyl chain rises the overall MPD of the IL and increases the refractive index. It is the case, for instance, of imidazolium-based cations combined with $[BF_4]^-$, $[PF_6]^-$ or $[NTf_2]^-$ anions. Oppositely, when the combination has higher MPD than that of the -CH_2- units, increasing the alkyl chain length diminishes the refractive index, as it happens in imidazolium-based cations paired with alkyl sulfates or dicyanamide anions.

In 2016, Bernardes et al. [30] published a continuation of the work of Bica addressing its limitations in predicting the refractive index of dicyanamide-based ILs. According to the authors, the worse performance of their model with these anions was because carbon hybridization was not taken into account, cursive, all carbons were assigned with the same electronic polarizability. Hence, in this new work, the authors expanded the set of ILs upon study beyond imidazolium-based combinations, and allowed the designed regression to distinguish between different types of carbon hybridization (sp, sp^2 and sp^3). The result was a model which was able to predict the refractive index at λ_D and the mass densities of ILs more accurately than that of Bica, with good performance on dicyanamide anions and pyridinium cations. It is worth to mention that the authors also checked the effect of considering hybridization in N and S atoms. However, they found no evidence of improvement in the refractive index prediction while the regression analysis became more complex. They also compared the atomic polarizability values derived from this model with those predicted by quantum mechanical calculations at the levels of theory MP2/augcc-pVDZ and CCSD(T)/aug-cc-pVDZ. The results revealed that their designed regression overestimates the polarizability in cations and underestimates it in anions because of the lacking/extra electron in those types of ions. However, since ILs are composed by both anions and cations, these errors counterbalance and the calculation of electronic polarizability in ionic pairs is unaffected.

In 2018, Heid et al. [31] explored the possibility of designing ILs with refractive indices higher than 2.0 RIU by using polyhalide anions. For doing it, the authors performed a new designed regression to include halides in the set of atomic polarizabilities derived by Bernardes et al. In their regression, they employed experimental refractive indices at λ_D of 55 halide-containing ILs while keeping constant the polarizabilities of the atoms already derived by Bernardes. In other words, the only fitting parameters were the polarizabilities of the halide atoms, Cl, Br, and I. The validity of the derived polarizabilities was successfully tested against quantum mechanical calculations at the M06-2X/Sadlej level of theory. The authors concluded that the MPD of iodide anions is the highest one among the considered halides, which means that iodide-based polyhalide anions are the best option to increase the value of refractive index. Indeed, they predicted that the refractive index of ILs based on combinations of common cations such as ammonium or imidazolium with polyiodides can easily go beyond 2.0 RIU.

More recently in 2020, Rodríguez-Fernández et al. [25] employed a semi-empirical approach to predict the refractive index in ILs as a function of both wavelength and temperature. The key idea is exploiting that the thermal and dispersive contributions to refractive index are independent in the Lorentz-Lorenz equation, Eq. (6).

The method requires knowing the experimental density of the IL at the desired temperature (molar volume, thermal dependence) and computing its wavelength-dependent electronic polarizability (dispersion). Specifically, the calculations in this work were done using the Coupled Perturbed Kohn-Sham (CPKS) method at the B3LYP/6-31++G(d,p) level of theory. The performance of this approach was evaluated comparing the predictions of the model with experimental data of 1-alkyl-3-methylimidazolium tetrafluoroborate-based ILs. The simulated refractive indices presented a relative root-mean square deviation of 0.4% in the thermal range from 293 to 313 K and the spectral range from 300 to 1500 nm. Hence, the authors confirmed that predictive models based on splitting the thermal and dispersive contributions to refractive index are able to provide accurate values for this magnitude.

In 2022, Rodríguez-Fernández et al. [32] published a new model for the fast prediction of the refractive index in ILs together with an atlas of the expected values for more than 1000 ion combinations. The method was developed to obtain accurate refractive indices at T = 298 K and λ = 589 nm by means of *ab initio* calculations of electronic polarizability (CPKS method) and molecular volumes on isolated ions. According to the authors, they were able to predict the refractive index of the 75% of the compounds in their test set (71 ILs, including imidazoliums, piperidiniums, pyrrolidiniums, pyridiniums, etc.) with an absolute error below 0.02 RIU. The level of theory employed was the B3LYP/6-31++G(d,p), chosen after testing other possibilities because its good performance/computational cost ratio.

In their model, electronic polarizability was additive in terms of ions, $\alpha_{IL} = \alpha_{anion} + \alpha_{cation}$, but not molecular volume, which requires a more delicate treatment given by: $V_{IL} = f_{scale}(V_{cation} + V_{anion}) + f_{int}(V_{cation} + V_{anion})^2$. In this expression, the f-terms are empirical parameters obtained by comparing experimental molecular volumes and *ab initio* predictions. f_{scale} mimics the effect of temperature and cavity volume in bulk ILs (DFT is performed at 0 K and in gas-phase), and f_{int} represents the packing ability and interactions of the ions. Note that the volume correction can be applied to single ions (a counterion is not required), hence, the MDP expected for isolated ionic species ($MPD = \alpha_{mol}/V_{mol}$) is also affected by this correction.

Using this volume correction, the authors showed that the species presenting charge delocalization tend to have largest MPD values than those which are aliphatic. In order to shed some light on this trend, they performed electronic polarizability calculations on an imidazolium cation with a conjugated chain of variable length. The simulations showed that electronic polarizability grows faster than expected from the linear addition of atomic polarizabilities in the chain, i.e., it becomes nonlinear. This finding is closely related to the effect of carbon hybridization in electronic polarizability shown by Bernardes et al., since sp^2 carbons in ILs often appear as a part of conjugated molecular regions. Last but not least, the authors pointed out that tuning charge delocalization in IL is a powerful tool to obtain liquids with refractive indices beyond 2.0 RIU.

Also in 2022, Koutsoumpos et al. [33] published a work aimed to predict the refractive index (or the mass density) of ILs by combining the electronic polarizability model of Wildman-Crippen and the Lorentz-Lorenz equation. The Wildman-Crippen model [34] was published in 1999 and it is based on assuming that electronic polarizability is an additive magnitude in terms of atomic contributions that are sensitive to aromaticity and chemical environment. Koutsoumpos et al. showed that using the atomic polarizabilities of Wildman-Crippen and combining them with experimental density measurements, the refractive index of ILs can be estimated with absolute errors as low as 0.002 RIU.

3.3 Artificial intelligence applied to refractive index prediction

In the realm of predicting the refractive index and its variations with different magnitudes, a recent and compelling approach involves the use of Artificial Intelligence (AI) learning systems. These methods have garnered significant attention, particularly with the development of specialized accelerators in various fields. The study of ILs is no exception, and, in recent years, there has been a proliferation of articles focused on the application of different AI techniques for analysis. This type of investigation closely aligns with semi-empirical models in terms of addressing the problem by using known magnitudes (such as chemical structure or other thermophysical properties) to predict the refractive index of materials.

In the context of predicting the refractive index of ILs, AI techniques, including machine learning models, have been employed to establish correlations between the molecular structure and optical properties. This involves training models on datasets containing information about the molecular features of various ILs and their corresponding refractive indices. Researchers frequently leverage computational chemistry methods, molecular descriptors, and advanced machine learning algorithms to construct accurate predictive models. These models contribute to a deeper understanding of the structure-property relationships that govern the optical behavior of ILs, thereby facilitating the design of innovative compounds with customized optical characteristics. They can be broadly categorized into two main groups: unsupervised models, which focus on identifying Principal Components (PCs) and reducing dimensionality for predictions, and supervised models, which predict the refractive index based on extensive databases.

While unsupervised learning is less commonly employed in predicting specific properties such as refractive index, it still finds utility in the pre-processing and exploratory phases of analysis, contributing to a comprehensive understanding of ILs. Unsupervised clustering techniques, including k-means clustering, hierarchical clustering, and density-based clustering, can be applied to unveil patterns or groups within the data. In addition, techniques such as Principal Component Analysis (PCA) or t-distributed stochastic neighbor embedding (t-SNE) serve to reduce data dimensionality, enhance visualization, and potentially optimize subsequent modeling efforts.

A notable example of the application of unsupervised models in the analysis of ILs is presented by the work of Venkatraman et al. [35], where researchers explore the main contributors influencing the refractive index using four different optimization routines available in the statistical software R. Tree-based approaches are reported to yield the best results, exhibiting a mean absolute error lower than 0.01 RIU for a set of 14 ILs.

Despite the effective performance of these models, the identification of the main contributions proves to be challenging. Venkatraman et al. addressed this task through PCA. PCs represent linear combinations of initial inputs, emphasizing the variables, such as temperature, with the greatest influence in the variation of the refractive index. The orthogonality of these PCs ensures minimal covariance, thereby reducing dimensionality. In the aforementioned work, the first two PCs accounted for 50% of the refractive index variance, demonstrating a successful reduction in complexity. Notably, these two PCs are predominantly influenced by descriptors related to the anion and cation, encompassing factors such as geometry, energy levels, gauge, and core-core repulsion. This approach demonstrates the capability of unsupervised models to classify ILs without prior knowledge and streamline the system into a more manageable PCs problem.

Beyond classification, the authors analyzed the influential variables on the refractive index using four different supervised routines (PLSR, GBM, Cubist, and RF). Notably, the models with superior system descriptions (Cubist and RF) highlight temperature as a crucial variable affecting refractive index variance. These studies contribute to narrowing the scope of dispersion studies of ILs by benchmarking their chemical and structural properties. However, it is essential to note that the absolute errors observed, even in the best models, are close to 0.01 RIU, suggesting the need for further refinement in predictive modeling.

In contrast to the previously discussed unsupervised models, supervised models predict the refractive index or polarizability of ILs based on molecular structure. These approaches are commonly known as Quantitative Structure-Property Relationship (QSPR) models, where "structure" typically refers to the arrangement of atoms in a molecule, while "property" encompasses specific chemical or physical characteristics, such as boiling point, solubility, toxicity, or refractive index.

Within this category, machine learning algorithms (Random Forests, Decision Trees, SVM, Neural Networks) are widely employed in ILs because of the descriptors used. Supervised methods require a labeled dataset with input features (descriptors) and corresponding target values (refractive indices) for training. However, a significant challenge arises from the unbalanced nature of these datasets, as specific ILs, usually those commercially available, are extensively studied, whereas others are only sporadically addressed. As depicted in **Figure 2** from Esmaeili et al. [36], the experimental data available is largely dominated by three cations (~80%) and a notable prevalence of the $[NTf_2]^-$ anion (~28%).

In the context of this approach, recent studies by Esmaeili et al. [36] and Baskin et al. [37] delve deeply into multiple widely used machine learning algorithm models and compare their utility. These studies utilize thousands of data points to develop models and conduct performance tests for various routines. In these cases, the input variables are not PCs calculated from unsupervised algorithms but rely on theoretical

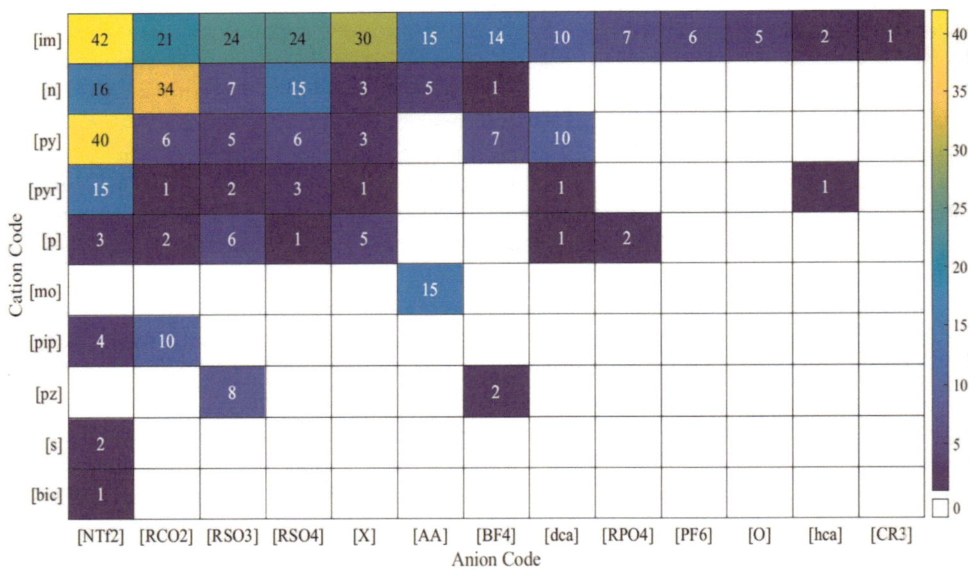

Figure 2.
Number of experimental entries employed by Esmaeili et al. for each type of IL as a function of their composing cations and anions. The figure was extracted from their work [36].

models based on the chemical structure of molecules, similar to those from Ref. [34]. Esmaeili et al. achieved outstanding performance in some routines (R^2 = 0.9973 for CatBoost), even incorporating wavelength-dependent values, although some conclusions remain debatable.

Two key insights emerge from these works that contradict previous studies. First, the influence of temperature and wavelength on the refractive index is downplayed in favor of structural variances, whereas PCA demonstrated a clear dominance of temperature over other chemical variances. The debate on this issue arises because of the wide range of thermo-optic coefficients and chromatic variations observed in ILs (refer to the experimental section). Second, the increase of the alkyl chain in imidazolium-based ILs is considered to always produce a increase in the refractive index. This contradicts findings discussed in studies such as those from Ref. [29], indicating that the refractive index can decrease with increasing alkyl length for a given anion in imidazolium-based liquids.

In a broad perspective, we can optimistically view AI as a powerful tool for refractive index analysis. Its synergy with computational models may highlight specific features crucial for refractive index tunability and enable reasonable prediction of their values. However, there is a long road ahead, with many analyses showing a non-negligible survivorship bias due to the concentration of experimental data on a limited set of liquids. Furthermore, the lack of experimental values for chromatic dispersion poses a challenge to these models, significantly limiting their performance.

4. Applications

Ionic materials, especially liquids are excellent candidates to be part of a new generation of photonic materials [38], with their recognized tunability being highly valued. Although the optical properties of ILs have been studied for the last 20 years, the most important advances at the fundamental and applied level have been made in the last decade, although there is still a long way to go to technologically consolidate them in the field of photonics.

Early published works presented ILs as potentially efficient materials for incorporation into devices operating in extreme climatic conditions. For example, they have been proposed as the basis of liquid mirrors for lunar telescopes [39] due to their low vapor pressure and their ability to remain liquid at temperatures as low as 130 K; or as optical thermometers [40] because they do not evaporate at temperatures reached in extremely hot regions such as some deserts.

Their potential application in optical mineralogy became apparent very early [41] as an alternative to the currently available high-index immersion fluids, which are volatile, toxic, or both. Recently, they also have been proposed as immersion liquids for filling porous materials, such as photonic crystals. In particular, in Ref. [19], ILs with high ($n_D > 1.7$ RIU) and low ($n_D < 1.5$ RIU) refractive indices were employed as immersion liquids to tune the optical properties of photonic crystals or to tune the refractive indices of the materials constituting these photonic crystals. Another remarkable implementation is the one of Guo et al. in Ref. [42]. They introduced an IL in a microstructured optical fiber to perform all-optical control of light transmission. For doing it, they exploited the dependence of the refractive index of the IL on temperature to modify the bandgap of the fiber and to control the attenuation through the system.

Subsequently, taking advantage of the influence of the refractive index on parameters such as group index or group velocity dispersion, Abood et al. [43] designed and computationally optimized a *Coupled-Slot Slab Photonic Crystal Waveguide*-based device to obtain slow light guided modes by filling parts of the structure with polymerized ILs of different refractive indices (1.3–2.1 RIU). The final device had high group index (low group velocity), a broad bandwidth and a wide spectral range with zero dispersion. The infiltration of the IL in the optimized device (for n = 1.8 RIU) improved its performance by reducing the bit length from 4.6748 to 4.2817 μm, increasing the bandwidth from 102.78 to 126.09 nm, the normalized delay bandwidth product from 0.663 to 0.724, the bit rate from 3.2 to 3.93 Tb/s and the average buffer capacity from 213.912 bits to 233.553 bits. Conversely, the group rate was reduced from 10.3 to 8.9 and delay times from 34.66 to 28.8 ps.

Sensors incorporating ILs in their composition have also been developed in recent years. In Ref. [44], a capacitive-type pressure sensor consisting in a porous polyvinylidene fluoride (PVDF) film sandwiched between two transparent electrodes is proposed. By filling the pores with an IL of the same refractive index as the PVDF, the transmittance of the film increases dramatically from 0% to 94.8% in the visible range. In addition to optical adaptation, the IL also significantly improves signal intensity and sensitivity due to the formation of an electrical double layer at the dielectric-electrode interfaces. Furthermore, it also enhances the hardness and elasticity of the active material due to a plasticizing effect. This sensor exhibits high flexibility, 90.4% transparency, a sensitivity of 1.19 kPa^{-1} and a low detection limit of ≈0.4 Pa, combining well the properties of flexibility, transparency, and detection.

Liquid-core fiber optic sensors filled with [C_2C_1im][Ac] and [C_4C_1im][Ac] have also been implemented to evaluate their suitability for CO_2 detection [45]. The CO_2 absorption in the liquids manifested in a reduction of refractive index (very similar in both ILs) and transmittance (attributed to the existence of scattering generated by the CO_2 bubbles).

Other field where the knowledge of the refractive index dispersion of ILs is essential is optofluidics, since it is required to design achromatic liquid lenses or dispersive prisms. Early research in this field developed variable focal liquid lenses [15, 46–48]. In these works, the focal length of the lens is controlled with pressure, electric field or by changing the filling liquid. Hu et al. used electrowetting to vary the curvature of the meniscus formed at the interface between two immiscible liquids, in this case an IL and dodecane. They tested ILs of the imidazolium family that span a refractive index range between 1.4089 and 1.5329 RIU: [C_2C_1im][ClO_4], [C_4C_1im][ClO_4], [C_4Py][BF_4], [C_2C_1im][NTf_2], [C_2C_1im][$N(CN)_2$], [C_4C_1im][OTf], [C_4C_1im][NO_3], [C_2C_1im][BF_4], and [C_4C_1im][PF_6]. Varying the applied voltage from 0 to 95 V, they obtained different focal distances from −1.5 to +1.5 m using the aforementioned liquids. To achieve significant focal changes, voltages above 60 V had to be applied.

Continuing with the fabrication of lenses, Calixto et al. [46] proposed the use of ILs for producing liquid microlenses in a polymer matrix. They used [C_2C_1im][BF_4], [C_4C_1im] [BF_4], [C_2C_1im][NTf_2], [C_4C_1im][NTf_2], [C_2C_1im][CF_3COO], and [C_4C_1im][CF_3COO] mixed with other solvents to achieve a wide range of refractive index variation and, hence, focal lengths. A few years later, Calixto et al. [15] designed an achromatic varifocal multichamber lens based on the three-chamber design of Reichelet and Zappe. For this purpose, they used ILs with high and low Abbe number. The focal length is modified by varying the amount of liquid/pressure in the chambers. Authors in Ref. [49] proposed a similar structure that also allowed the size of the two chambers to be varied by injecting or withdrawing liquid, simplifying the zoom function and minimizing aberrations.

There are also studies exploring the use of ILs in electro-optical devices by investigating the variation of the refractive index and the appearance of birefringence when an electric field is applied. In Ref. [50], a little more knowledge is advanced by trying to understand the mechanisms that lead to the modification of the refractive index in the transverse direction to the application of the field and, consequently, to the variation of the focal length of a lens formed by the IL $[C_8C_1im][NTf_2]$. The authors related the refractive index gradient to modest changes that occur in the surface charge density, resulting in a free charge density gradient in the liquid. They estimated the thermal effects on the refractive index generated by the Joule effect, whose contribution is much smaller than that of the applied field.

Signal modulation in integrated photonic devices requires optical phase control that can be exercised by manipulating the effective refractive index of the waveguide. Different strategies have been considered for the design of SiN-based optical modulators. In Ref. [49], it is proposed the fabrication of a ring microresonator consisting of a MoS layer structure on a SiN substrate coated by the IL $[DEME][NTf_2]$ to control the charge doping processes. Direct and inverse bias voltages change the effective index of the waveguide, producing a change in the transmission spectra due to a redshift or a blueshift of the resonance wavelength.

The refractive index is a property sensitive to enviromental factors such as humidity, which induces water absorption in some ILs. Therefore, variations in the refractive index can be used to estimate the amount of water absorbed by an IL. In Ref. [51], a neural network-based model was developed to estimate the purity of an IL from the molecular mass, the nominal refractive index and the measured refractive index. It was applied to the ILs $[C_4C_1im][BF_4]$ and $[C_4C_1im][C_1SO_4]$. Measurable differences in refractive index were obtained when the water content was greater than 1% by weight.

On the other hand, Kaneko et al. [52] studied with a simpler model the effects of the presence of a known amount of water on the refractive index. An indirect application of their study is the determination of the water mole fraction from the values of the dry IL and the IL-water mixture. It is worth to note that most standardized refractive index measurement systems do not usually have sufficient sensitivity to detect the presence of impurities or very small amounts of water, being necessary to resort to more sensitive techniques such as surface plasmon resonance sensors that work well detecting refractive index variations that can be of the order of 10^{-5} or 10^{-6} RIU.

5. Conclusions

The initial measurements of optical properties in ILs aimed to provide a comprehensive physicochemical characterization of these emerging materials. As the number of experimental measurements increased, namely those of the refractive index at the sodium D line, the possibility of modifying their optical response through structural changes became apparent. This goal has been the focus of several recent experimental and computational studies. From an experimental perspective, refractive index measurements have been expanded beyond single-wavelength to include wider spectral regions, allowing for the study of its wavelength dependence in the visible, ultraviolet, and infrared ranges. In consequence, these studies have investigated the impact of different ion combinations on refractive index, not only at the sodium D line, but also across the entire refractive index curve. From a computational perspective, many publications have focused on predicting the refractive index of specific ion combinations, which lays the groundwork for future experimental work. The computational efforts were based on two different approaches. The first one exploits the additivity of electronic polarizability in terms of atoms or molecular units to build the overall electronic polarizability of new ILs and their refractive index via the Lorentz-Lorenz equation. The other approach involves applying artificial intelligence algorithms to large amounts of experimental data to create models with the ability to predict the refractive index of new ILs. These experimental and computational efforts to unravel the relationships between molecular structure and refractive index in ILs, has led to an increase in the number of works investigating their potential optical applications. As a result of it, ILs are being studied as optical materials in photonic fields such as optical communications and sensing, in addition to early-proposed applications such as high refractive index immersion fluids or variable focal length liquid lenses.

Abbreviations

Cation	
$[C_kC_1im]^+$	1-alkyl-3-methylimidazolium
$[BzC_1im]^+$	1-benzyl-3-methylimidazolium
$[AllC_1im]^+$	1-allyl-3-methylimidazolium
$[AllC_1C_1im]^+$	1-allyl-2,3-dimethylimidazolium
$[C_4C_1imI_2]^+$	4,5-diiodo-1-butyl-3-methylimidazolium
$[C_2CNC_4im]^+$	3-(3-butyl-1H-imidazol-3-ium-1-yl)propanenitrile
$[C_2CNAim]^+$	3-(3-allyl-1H-imidazol-3-ium-1-yl)propanenitrile
$[C_2CNBzim]^+$	3-(3-benzyl-1H-imidazol-3-ium-1-yl) propanenitrile
$[C_2CNHeim]^+$	3-[3-(2-hydroxyethyl)-1H-imidazol-3-ium-1-yl]propanenitrile
$[C_2OC_2C_1im]^+$	1-ethoxyethyl-3-methylimidazolium
$[C_4(C_1im)_2]^{2+}$	1,1-bis(3-methylimidazolium-1-yl) butylene
$[N_{kk}]^+$	di(alkyl)ammonium
$[N_{4444}]^+$	tetra(butyl)ammonium
$[DEME]^+$	diethylmethyl(2-methoxyethyl)ammonium
$[P_{444k}]^+$	tri(butyl)-alkyl-phosphonium
$[P_{kkkk}]^+$	tetra(alkyl)phosphonium
$[P_{666,14}]^+$	tri(hexyl)tetradecylphosphonium
$[P_{ppp6}]^+$	hexyltryphenyl-phosphonium
$[P_{i(444)1}]^+$	tri(isobutyl) methylphosphonium
$[C_1py]^+$	1-methylpyridinium
$[piC_5py]^{2+}$	1-(1-methypiperidinium-1-yl) pentane-(1-pyridinium)
$[Ch]^+$	cholinium
$[C_1C_kMor]^+$	N-alkyl-N-methylmorpholinium
Anion	
$[I_5]^-$	pentaiodide
$[PF_6]^-$	hexafluorophosphate
$[BrI_2]^-$	bromodiiodide
$[HSO_4]^-$	hydrogensulfate
$[C_1SO_3]^-$	methanesulfonate
$[C_1SO_4]^-$	methylsulfate
$[Bz\text{-}I\text{-}SO_3]^-$	2-iodobenzenesulfonate
$[OTf]^-$	trifluoromethanesulfonate

$[NTf_2]^-$	bis(trifluoromethylsulfonyl)imide
$[SCN]^-$	thiocyanate
$[MOPSO]^-$	2-hydroxy-3-morpholinopropanesulfonate
$[BES]^-$	2-(bis(2-hydroxyethyl)amino)ethanesulfonate
$[TAPSO]^-$	N-[tris(hydroxymethyl)methyl]-3-amino-2-hydroxypropane-sulfonate
$[CAPSO]^-$	3-(cyclohexylamino)-2-hydroxypropanesulfonate
$[\rho\text{-}TSA]^-$	paratoluene sulfonate
$[(R\text{-}O)_2PO_2]^-$	di(alkyl)phosphate
$[C_1C_1OPO_2]^-$	methyl-methylphosphonate
$[C_1OHPO_2]^-$	methylphosphonate
$[Ac]^-$	acetate
$[Ala]^-$	N-acetyl-L-alaninate
$[Val]^-$	N-acetyl-L-valinate
$[Leu]^-$	N-acetyl-L-leucinate
$[Ile]^-$	N-acetyl-L-isoleucinate
$[C_kCO_2]^-$	alkanoate
$[CF_3COO]^-$	trifluoroacetic acid
$[Tos]^-$	tosylate
$[Sac]^-$	saccharinate
$[N(CN)_2]^-$	dicyanamide

Author details

Yago Arosa[1], Carlos Damián Rodríguez-Fernández[2], Elena López Lago[1] and Raúl De la Fuente[1]*

1 Departamento de Física Aplicada e Instituto de Materiais (iMATUS), Universidade de Santiago de Compostela, Spain

2 Departamento de Física e Ciencias da Terra, Universidade da Coruña, Spain

*Address all correspondence to: raul.delafuente@usc.es

References

[1] Bonhôte P, Dias A-P, Papageorgiou N, Kalyanasundaram K, Grätzel M. Hydrophobic, highly conductive ambient-temperature molten salts. Inorganic Chemistry. 1996;**35**(5): 1168-1178

[2] Huddleston JG, Visser AE, Reichert WM, Willauer HD, Broker GA, Rogers RD. Characterization and comparison of hydrophilic and hydrophobic room temperature ionic liquids incorporating the imidazolium cation. Green Chemistry. 2001;**3**(4): 156-164

[3] De Los Ríos AP, Fernández FJH. Ionic Liquids in Separation Technology. Amsterdam, Netherlands: Elsevier; 2014

[4] Kayama Y, Ichikawa T, Ohno H. Transparent and colourless room temperature ionic liquids having high refractive index over 1.60. Chemical Communications. 2014;**50**(94): 14790-14792

[5] Gonfa G, Bustam MA, Muhammad N, Khan AS. Evaluation of thermophysical properties of functionalized imidazolium thiocyanate based ionic liquids. Industrial and Engineering Chemistry Research. 2015;**54**(49):12428-12437

[6] Ullah S, Al-Sehemi AG, Assiri MA, Mukhtar A, Ayoub M, Bustamb MA, et al. Experimental investigation and modeling of the density, refractive index, and dynamic viscosity of 1-propyronitrile-3-butylimidazolium dicyanamide. Journal of Molecular Liquids. 2020;**302**:112470

[7] Bhattacharjee A, Lopes-da-Silva JA, Freire MG, Coutinho JAP, Carvalho PJ. Thermophysical properties of phosphonium-based ionic liquids. Fluid Phase Equilibria. 2015;**400**:103-113

[8] Rodil E, Arce A, Arce A, Soto A. Measurements of the density, refractive index, electrical conductivity, thermal conductivity and dynamic viscosity for tributylmethylphosphonium and methylsulfate based ionic liquids. Thermochimica Acta. 2018;**664**:81-90

[9] Othman Zailani NHZ, Yunus NM, Ab Rahim AH, Bustam MA. Thermophysical properties of newly synthesized ammonium-based protic ionic liquids: Effect of temperature, anion and alkyl chain length. Processes. 2020;**8**(6):742

[10] Lee SY, Vicente FA, Coutinho JAP, Khoiroh I, Show PL, Ventura SPM. Densities, viscosities, and refractive indexes of good's buffer ionic liquids. Journal of Chemical & Engineering Data. 2016;**61**(7):2260-2268

[11] Marcinkowski Ł, Szepiński E, Milewska MJ, Kloskowski A. Density, sound velocity, viscosity, and refractive index of new morpholinium ionic liquids with amino acid-based anions: Effect of temperature, alkyl chain length, and anion. Journal of Molecular Liquids. 2019;**284**:557-568

[12] Khan AS, Man Z, Arvina A, Bustama MA, Nasrullah A, Ullah Z, et al. Dicationic imidazolium based ionic liquids: Synthesis and properties. Journal of Molecular Liquids. 2017;**227**:98-105

[13] Branco ASH, Calado MS, Fareleira JMNA, Visak ZP, Canongia Lopes JN. Refraction index and molar refraction in ionic liquid/PEG200 solutions. Journal of Solution Chemistry. 2015;**44**:431-439

[14] Wang J, Song H, Yang X, Zou W, Chen Y, Duan S, et al. Density, viscosity, refraction index and excess properties of

binary mixtures of 1-(1-methypiperidinium-1-yl) pentane-(1-pyridinium) bis(trifluoromethane) sulfonamide with acetonitrile at T= (293.15 to 323.15) K. Korean Journal of Chemical Engineering. 2016;**33**(8): 2460-2468

[15] Calixto S, Rosete-Aguilar M, Sanchez-Marin FJ, Torres-Rocha OL, Prado EMM, Calixto-Solano M. Optofluidic compound lenses made with ionic liquids. In: Applications of Ionic Liquids in Science and Technology. London, UK: InTech; 2011

[16] Nóvoa-López JA, López Lago E, Domínguez-Pérez M, Troncoso J, Varela LM, de la Fuente R, et al. Thermal refraction in ionic liquids induced by a train of femtosecond laser pulses. Optics and Laser Technology. 2014;**61**:1-7

[17] Chiappe C, Margari P, Mezzetta A, Pomelli CS, Koutsoumpos S, Papamichael M, et al. Temperature effects on the viscosity and the wavelength-dependent refractive index of imidazolium-based ionic liquids with a phosphorus-containing anion. Physical Chemistry Chemical Physics. 2017; **19**(12):8201-8209

[18] Mero A, Guglielmero L, D'Andrea F, Pomelli CS, Guazzelli L, Koutsoumpos S, et al. Influence of the cation partner on levulinate ionic liquids properties. Journal of Molecular Liquids. 2022;**354**: 118850

[19] Wu X, Muntzeck M, de los Arcos T, Grundmeier G, Wilhelm R, Wagner T. Determination of the refractive indices of ionic liquids by ellipsometry, and their application as immersion liquids. Applied Optics. 2018;**57**(31):9215

[20] Rola K, Zajac A, Czajkowski M, Fiedot-Tobola M, Szpecht A, Cybinska J, et al. Electron beam patterning of polymerizable ionic liquid films for application in photonics. Langmuir. 2019;**35**(37):11968-11978

[21] Elmahdy MM, Fahmy T, Aldhafeeri KA, Ibnouf EO, Riadi Y. Optical and antibacterial properties of 1-butyl-3-methylimidazolium ionic liquids with trifluoromethanesulfonate or tetrafluoroborate anion. Materials Chemistry and Physics. 2021;**264**:124369

[22] Arosa Y, Rodríguez-Fernández CD, López Lago E, Amigo A, Varela LM, Cabeza O, et al. Refractive index measurement of imidazolium based ionic liquids in the Vis-NIR. Optical Materials. 2017;**73**:647-657

[23] Arosa Y, Algnamat BS, Rodríguez CD, López Lago E, Varela LM, de la Fuente R. Modeling the temperature-dependent material dispersion of imidazolium-based ionic liquids in the VIS-NIR. Journal of Physical Chemistry C. 2018;**122**(51):29470-29478

[24] Algnamat BS, Arosa Y, López Lago E, de la Fuente R. An inspection of the dispersive properties of imidazolium-based ionic liquids in the Vis-NIR. Optical Materials. 2020;**102**: 109764

[25] Rodríguez Fernández CD, Arosa Y, Algnamat B, López Lago E, de la Fuente R. An experimental and computational study on the material dispersion of 1-alkyl-3-methylimidazolium tetrafluoroborate ionic liquids. Physical Chemistry Chemical Physics. 2020;**22**(25): 14061-14076

[26] Gardas RL, Coutinho JAP. Group contribution methods for the prediction of thermophysical and transport properties of ionic liquids. AICHE Journal. 2009;**55**(5):1274-1290

[27] Sattari M, Kamari A, Mohammadi AH, Ramjugernath D. A group contribution method for estimating the refractive indices of ionic liquids. Journal of Molecular Liquids. 2014;**200**:410-415

[28] Izgorodina EI, Forsyth M, MacFarlane DR. On the components of the dielectric constants of ionic liquids: Ionic polarization? Physical Chemistry Chemical Physics. 2009;**11**(14):2452

[29] Bica K, Deetlefs M, Schröder C, Seddon KR. Polarisabilities of alkylimidazolium ionic liquids. Physical Chemistry Chemical Physics. 2013;**15**(8): 2703

[30] Bernardes CES, Shimizu K, Canongia Lopes JN, Marquetand P, Heid E, Steinhauser O, et al. Additive polarizabilities in ionic liquids. Physical Chemistry Chemical Physics. 2016;**18**(3): 1665-1670

[31] Heid E, Heindl M, Dienstl P, Schröder C. Additive polarizabilities of halides in ionic liquids and organic solvents. The Journal of Chemical Physics. 2018;**149**(4):044302

[32] Rodríguez-Fernández CD, López Lago E, Schröder C, Varela LM. Non-additive electronic polarizabilities of ionic liquids: Charge delocalization effects. Journal of Molecular Liquids. 2022;**346**:117099

[33] Koutsoumpos S, Chronaki M, Tsonos C, Karakasidis T, Guazzelli L, Mezzetta A, et al. On the application of the Wildman-Crippen model to ionic liquids. Results in Materials. 2022;**16**: 100350

[34] Wildman SA, Crippen GM. Prediction of physicochemical parameters by atomic contributions. Journal of Chemical Information and Computer Sciences. 1999;**39**(5):868-873

[35] Venkatraman V, Raj JJ, Evjen S, Lethesh KC, Fiksdahl A. In silico prediction and experimental verification of ionic liquid refractive indices. Journal of Molecular Liquids. 2018;**264**: 563-570

[36] Esmaeili A, Hekmatmehr H, Atashrouz S, Madani SA, Pourmahdi M, Nedeljkovic D, et al. Insights into modeling refractive index of ionic liquids using chemical structure-based machine learning methods. Scientific Reports. 2023;**13**(1):1966

[37] Baskin I, Epshtein A, Ein-Eli Y. Benchmarking machine learning methods for modeling physical properties of ionic liquids. Journal of Molecular Liquids. 2022;**351**:118616

[38] Zhang S, Zhang Q, Zhang Y, Chen Z, Watanabe M, Deng Y. Beyond solvents and electrolytes: Ionic liquids-based advanced functional materials. Progress in Materials Science. 2016;**77**(6):80-124

[39] Borra EF, Seddiki O, Angel R, Eisenstein D, Hickson P, Seddon KR, et al. Deposition of metal films on an ionic liquid as a basis for a lunar telescope. Nature. 2007;**447**(7147): 979-981

[40] Rogers RD, Seddon KR. Ionic Liquids IIIB Fundamentals, Progress, Challenges and Opportunities. Transformations and process, ACS Sympos. Oxford, United Kingdom: Oxford University Press; 2005

[41] Deetlefs M, Shara M, Seddon KR. Refractive indices of ionic liquids. ACS Symposium Series. 2005;**901**:219-233

[42] Guo J, Zhou M, Liu YG, Di K, Li R, Cui W, et al. An all-optical controlled

attenuation effect in an all-fiber system based on ionic liquid-filled photonic bandgap fiber. Physica Scripta. 2019; **94**(11):115508

[43] Abood I, Elshahat S, Ouyang Z. High figure of merit optical buffering in coupled-slot slab photonic crystal waveguide with ionic liquid. Nanomaterials. 2020;**10**(9):1742

[44] Liu Q, Liu Z, Li C, Xie K, Zhu P, Shao B, et al. Highly transparent and flexible iontronic pressure sensors based on an opaque to transparent transition. Advancement of Science. 2020;**7**(10): 2000348

[45] Ohkura M, Takana H, Ohuchi FS, Furukawa R. Fabrication of liquid-core fiber-optic structure for large-area CO_2 sensing using ionic liquids. Journal of Fluid Science and Technology. 2021; **16**(1):JFST0004

[46] Calixto S, Sánchez-Morales ME, Sánchez-Marin FJ, Rosete-Aguilar M, Richa AM, Barrera-Rivera KA. Optofluidic variable focus lenses. Applied Optics. 2009;**48**(12):2308

[47] Hu X, Zhang S, Qu C, Zhang Q, Lu L, Ma X, et al. Ionic liquid based variable focus lenses. Soft Matter. 2011; **7**(13):5941

[48] Hu X, Zhang S, Liu Y, Qu C, Lu L, Ma X, et al. Electrowetting based infrared lens using ionic liquids. Applied Physics Letters. 2011;**99**(21):213505

[49] Chen H, Zhao Z, Zhang Z, Wang G, Li J, Shang Z, et al. Heterogeneous integrated phase modulator based on two-dimensional layered materials. Photonics Research. 2022;**10**(6):1401

[50] Wang Y, Swain GM, Blanchard GJ. Charge-induced birefringence in a room-temperature ionic liquid. The Journal of Physical Chemistry. B. 2021;**125**(3): 950-955

[51] Díaz-Rodríguez P, Cancilla JC, Matute G, Chicharro D, Torrecilla JS. Inputting molecular weights into a multilayer perceptron to estimate refractive indices of dialkylimidazolium-based ionic liquids—A purity evaluation. Applied Soft Computing. 2015;**28**: 394-399

[52] Kaneko K, Yoshimura Y, Shimizu A. Water concentration dependence of the refractive index of various ionic liquid-water mixtures. Journal of Molecular Liquids. 2018;**250**:283-286

Chapter 3

Applications of Ionic Liquids in Pharmaceuticals

Sadia Rani, Ali Irfan, Samina Aslam, Gamal A. Shazly, Huma Hameed, Muhammad Zaman, Noreen Samad and Yousef A. Bin Jardan

Abstract

Ionic liquids (ILs) have emerged as a cutting-edge frontier in the realm of science and technology, offering a unique blend of tunable physicochemical and biological features with eco-friendly characteristics. Thus, ILs exhibit a broad spectrum of applications in pharmacology, pharmaceuticals, medicine, and pharmaceutical fields, providing solutions to challenges in drug formulation and delivery. The focus of this chapter is on the application of IL methodologies and strategies to resolve critical issues within the pharmaceutical field, such as polymorphism, low solubility, stability, and bioavailability, which are problems with solid-state pharmaceuticals. The innovative use of ILs as carriers for active pharmaceutical ingredients (APIs) presents a highly promising avenue for addressing these challenges. In conclusion, this chapter detailed the utilization of ILs in pharmaceutical applications, and the strategic design of liquid salts has the potential to revolutionize the way we address critical issues in drug development, manufacturing, and developing biocompatible ILs.

Keywords: ionic liquids (ILs), ILs as solvent, ILs in drug delivery, ILs in drug formulations, API-ILs, ILs in drug extraction, toxicity & biodegradability of ILs

1. Introduction

The current challenging task for the pharmaceutical industries is the development of safe, effective and smart chemotherapeutic formulations, and delivery systems for the management of various diseases [1]. Pharmaceutical industry portfolios are facing an ever-increasing issue with the limited aqueous solubility of active pharmaceutical ingredients (APIs) in biopharmaceutical classification system class II drugs [2]. The difficulties encountered in the development of new drug formulations are mainly associated with the slow dissolution of APIs in biological fluids, which leads to insufficient and irregular systematic exposure, resulting in suboptimal clinical efficacy. The pharmaceutical companies are actively exploring various strategies to surmount these challenges, recognizing them as pivotal for enhancing drug performance. Some of the approaches under investigation include prodrug [2, 3], micellar system [4], salt formation [5], solid dispersions [6], and crystal engineering [7].

 IntechOpen

The pharmaceutical sector was initially motivated to investigate ILs due to their distinct characteristics and potential as environmentally friendly solvents. Ionic liquids (ILs) are defined as solvents with a 1:1 ratio of bulky, asymmetric cations and anions that are shaped by non-directional ionic interactions and exist as liquids below 100°C [8]. Other names for ILs include ionic glasses, ionic melts, designer solvents, fusion salts, green solvents, ionic fluids, liquid electrolytes, and solvents of the future [9].

These solvents belong to a unique class that is ideal for biological material formulation and development while also being environmentally benign. Paul Walden developed ethylammonium nitrate, the first ionic liquid, in 1914 [10]. Because of how difficult it was to handle and how sensitive it was to moisture and air, this initial ionic liquid was not employed much. This was altered in 1992 when Wilkes and Zaworotko provided the first set of imidazolium cation-based air- and moisture-stable ionic liquids [11]. Since then, these have found widespread application in a number of interdisciplinary domains, including biomedicine, engineering, and material chemistry [8, 12–14]. With the expansion of their chemical diversity, ILs have been further subdivided into numerous types, such as supported IL membranes (SILMs) [15] that comprise composites of ILs supported on metal–organic frameworks (MOFs) [16], task-specific ILs (TSILs) [17], room-temperature ILs (RTILs) [18], polyionic liquids (PILs) [19], and so on. It is frequently necessary for scientists and engineers to swiftly screen for appropriate ILs for a particular process. Ionic liquids often consist of inorganic anions and organic cations that include nitrogen [20]. **Figure 1** illustrates a few typical cation and anion structures [8, 21, 22].

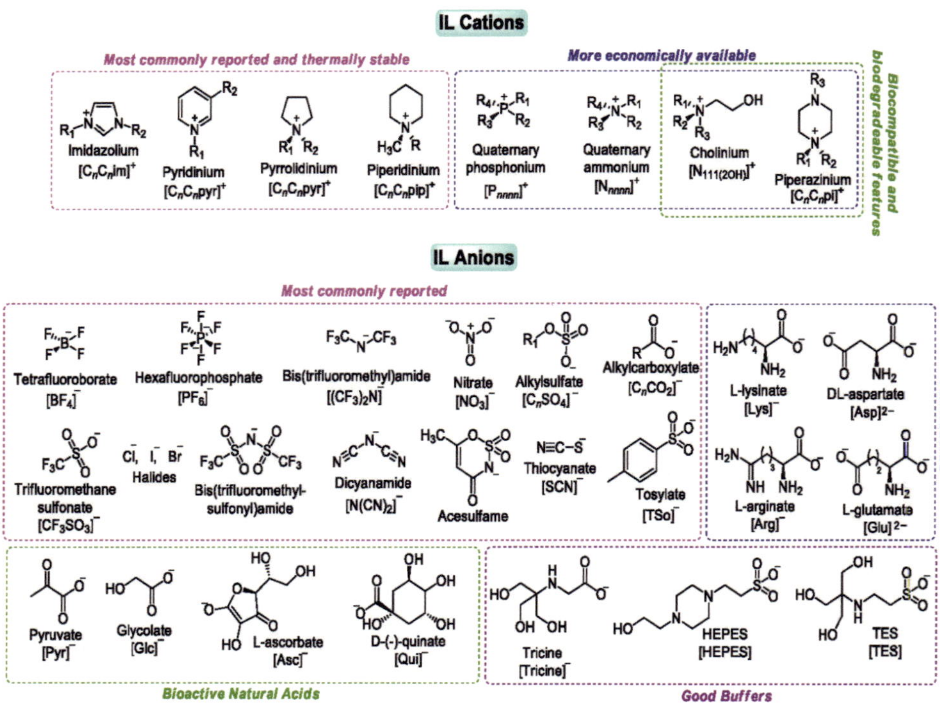

Figure 1.
Examples of IL cations and anions' chemical structures.

2. Evolution of ionic liquids

Four generations of ILs are currently on the market for different purpose (**Figure 2**) [1, 25].

2.1 Generation 1

Particular physical characteristics of the first generation of ILs include a low melting point, good thermal stability, reduced vapor pressure, and increased fluidity [26–28]. These ILs could take the place of organic solvents and agents that are flammable and bad for the environment. The majority of first-generation ILs, including $[C_4MIM][BF_4]$ and $[C_4MIM][PF_6]$, are aquatic ecotoxic and poorly biodegradable. But the first-generation ILs are also linked to additional weakly coordinating anions such as methylsulfate TFA, triflate, and TFSA [26, 27].

2.2 Generation 2

With their customizable physical and chemical characteristics, second-generation ILs are particularly useful in metal ion complexation, lubricants, and energetic materials [23, 26, 27]. These ILs have higher concentrations of biocompatible cations and anions that come from naturally occurring, reasonably priced, environmentally benign sources such as amino acids and carbohydrates. For instance, cholinium ILs can be made by reacting a number of naturally available carboxylic acids with cheap choline hydroxide to generate carboxylate salts and water as the only byproduct [27].

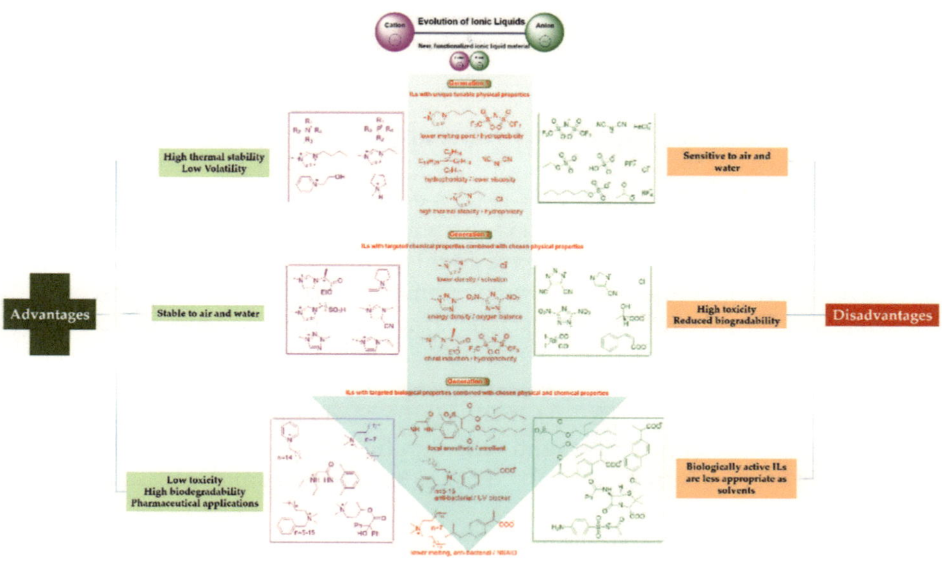

Figure 2.
Generations of ionic liquids: the evolution through time highlighting their characteristics, benefits, and drawbacks [23, 24].

2.3 Generation 3

Third-generation ILs have a biological feature added to their tunable chemical and physical properties, making them suitable for use in biological and pharmaceutical applications [23, 26, 29, 30]. Their suitability for biopharmaceutical applications, including local anesthetic, antifungal, antibacterial, and anticholinergic drugs, is mostly due to biologically active ions derived from recognized biocompatible sources. To fabricate various kinds of ILs, active pharmaceutical ingredients (APIs) or API precursors are primarily utilized as cations, anions, or both cations and anions [26, 29]. For biological uses, ILs in this group can also be easily biodegraded [31]. In comparison to the first- and second-generation ILs, the third-generation ILs also have some noteworthy advantages, including low production costs, an easy tunability of anions and cations, the structural and functional integrity of API, and controllable polymorphism [26].

In particular, the pharmaceutical industry is making more and more use of ILs because of their advantages in terms of drug delivery (oral, transdermal, and topical), stability, and solubility. Finally, by converting solid-state APIs into the IL form, potential issues with solid-state APIs can be easily overcome [32, 33]. Over the past 20 years, there has been a notable rise in the quantity of research publications concerning the incorporation of ILs in therapeutic formulations and administration [8, 34, 35]. In particular, the third generation of ILs has been used extensively to create ILs with desirable biological activity as active pharmaceutical ingredients (APIs) [36].

2.4 Generation 4

In 2018, the fourth generation of ionic liquids came out. These ionic liquids, whether in solution or after combining with other molecular liquids, had distinct and unanticipated properties and were biocompatible [25].

3. Applications of ionic liquids in pharmaceuticals

The primary areas of research for the next decade years will be the use of ILs in biomedicine [37]. In addition to their particular features and application as solvents, tunable ionic liquids (ILs) have enabled the emergence of novel, important topics. It is remarkable considering how many IL studies are now shifting their attention toward the biological sciences and medicine. Because of its unique attributes, the pharmaceutical industry is increasingly employing IL as a solvent [38]. Researchers can alter their characteristics to develop effective ILs that may be used in biomedical research. The uses of ILs in biomedicine have grown to include biosensors, drug delivery carriers, and other applications because of their structural tunability and biocompatibility [8, 10]. **Figure 3** shows examples of IL applications in each of these fields.

In order to increase a drug molecule's solubility, pharmacokinetic and pharmacodynamic capabilities, skin and intestinal permeability, formulation stability, and other attributes, ILs have been widely utilized in recent years to modify the physicochemical and biopharmaceutical properties of a therapeutic molecule [1, 8, 10]. The type of the cation and anion as well as the substituents on the cation or anion determine the biological and physicochemical properties of ILs.

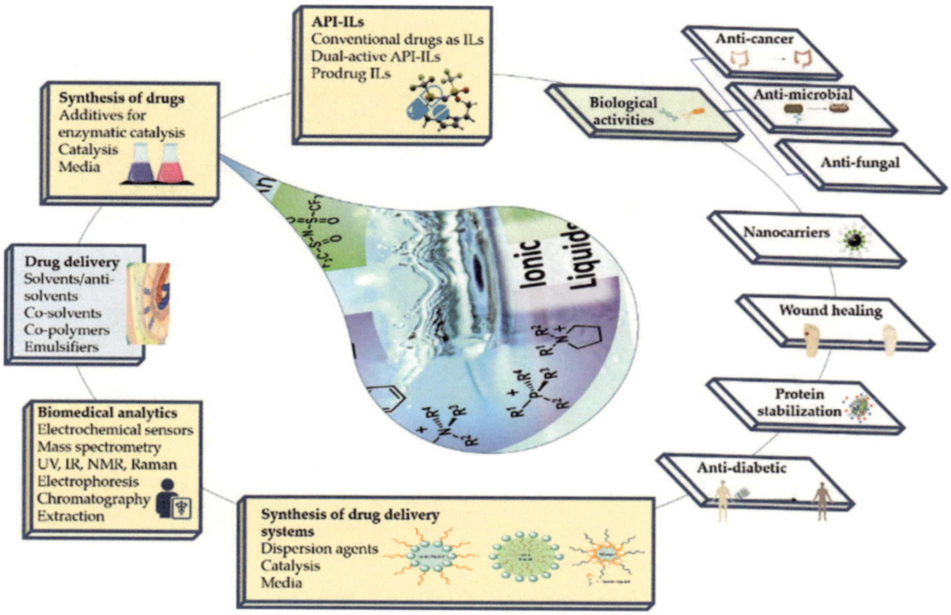

Figure 3.
Biomedical applications of ionic liquids in biomedicines.

Ionic liquids have been employed in biological applications since the late 1990s and early 2000s. They have been used to increase enzymatic catalytic efficiency and improve the thermostability of model proteins and enzymes (**Figure 4b**) [40, 41]. ILs were also employed as formulation excipients [42] for small compounds with poor water solubility and in controlled release systems [43]. Because of their basic chemical characteristics, which set them apart as particularly potent molecules for biomedical applications, ILs have gradually extended out into several domains over the years. Moreover, ILs have become a cutting-edge drug delivery platform [44–46]. ILs have the potential to increase the permeability of small and macromolecular drugs across biological barriers [10, 47], as well as to boost the solubility of poorly water-soluble pharmaceuticals by several to thousand folds [21]. Certain ILs are bioactive, showing antimicrobial (**Figure 5**) [49], anti-oxidant [8, 50], anti-cancer potential (**Figure 4a**) [51], anti-proliferative [52], anti-biofilm activities [53], and anti-diabetic (**Figure 6**) [54].

Levodopa is a widely used medicine for Parkinson's disease. ILs facilitated in the development of levodopa. IL was thought to be a more economical solvent than methanol, which was the typical solvent in the process but had the drawback of requiring higher pressure [1, 39]. Furthermore, ILs may be employed as a medium for the tailored synthesis of pharmaceutical nanocrystals. The use of nanocrystals in the delivery of poorly soluble drugs is very beneficial since it allows for improved bioavailability and solubility [55]. Ionic liquids were first employed in biomedical nanocarriers when they were used to synthesize and template silica nanoparticles for drug release [56], gene transfection [57], and antimicrobial applications. Subsequently, they were employed as emulsifiers [58] in nanocarriers as well as in nanocomposite systems [59] and the fabrication of nanoparticles using non-silica materials [60]. In order to enhance many of the properties of nanocarriers, such as biocompatibility,

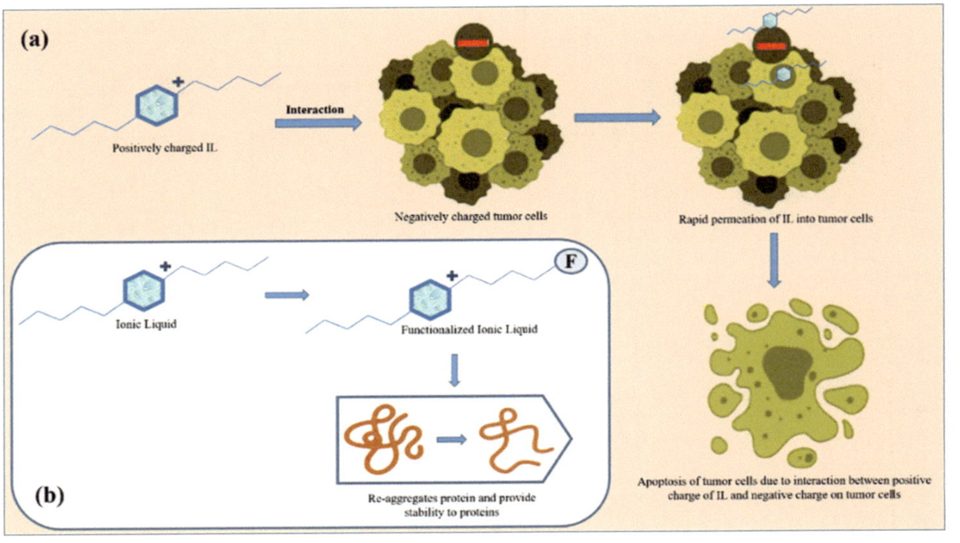

Figure 4.
(a) Schematic representation of ionic liquid demonstrating anticancer activity. (b) Diagram showing how the stability of proteins is affected by ionic liquid [39].

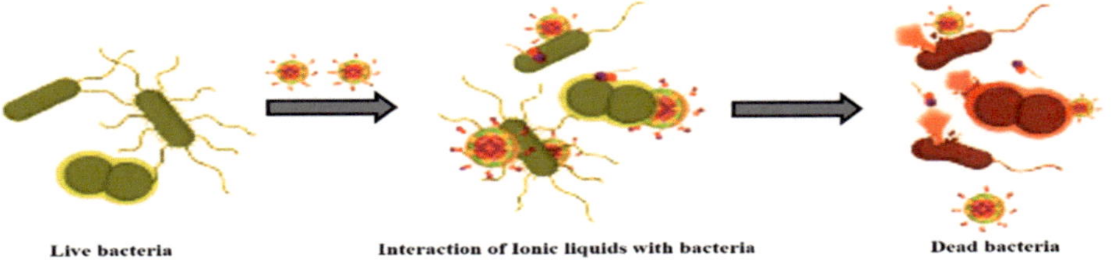

Figure 5.
Antibacterial activity of ionic liquids [48].

stability, and drug loading, ILs have also lately been utilized in their synthesis [61]. Use of ionic liquids in drug delivery system is displayed in **Table 1**.

3.1 Ionic liquids in drug synthesis

Ionic liquids in drug production have thereby started a new era. The use of ILs in a variety of chemical processes associated with the synthesis of medicines, their precursors or intermediates, and other substances with proven or potential biological activity has been documented by numerous research groups over the past 15 years. ILs are used in the synthesis of heterocyclic compounds that are used in biology and medicine, including imidazoles, furans, oxazoles, thiazoles, and quinolones [66]. These heterocyclic compounds are API's precursors with extensive biological activities, due to their high applicability in many chemical processes [67]. Ionic liquids have exciting applications in the manufacture of medicinal compounds and pharmaceuticals, among their many other uses in chemical processes. For example, the NSAID pravadoline was synthesized using $[C_4MIM][PF_6]$ as the media, signifying the first high-yield synthesis of a drug in an ionic liquid medium, as reported in 2000 [68]. By heating the IL to

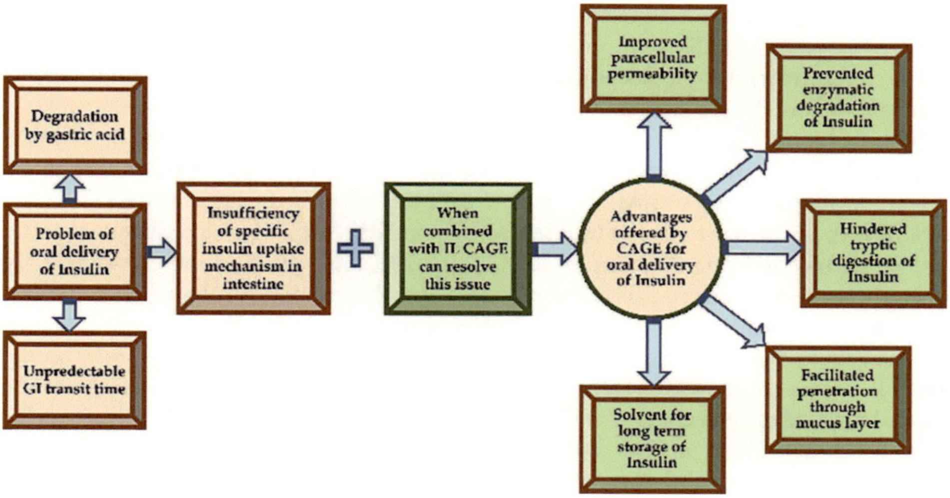

Figure 6.
An illustration showing the benefits of ionic liquid for diabetes [39].

IL used	Description	Application	Advantage	Ref.
Choline amino acid IL	The IL system boosted rutin's bioavailability and expanded its use in the treatment of cancer	To increase rutin's anticancer potential	Provided improved solubility of rutin (a poorly soluble drugs) and regulated drug delivery	[62]
1-Dodecyl-3-methylimidazolium bromide	One of Quercetin's problems is its solubility. The solubility issues can be resolved by surface active IL (SAILS). The non-ionic polymers' interaction with SAILs allowed for effective drug delivery	Anti-tumorous and anti-inflammatory actions	High anti-tumor effectiveness, improved hydrophobic drug solubility, and prolonged release	[63]
Choline and geranic acid	Enhancing the solubility of diabetes medications is crucial for augmenting their bioavailability. The CAGE IL system increased nobiletin's solubility	Antidiabetic potential	Enhanced drug permeability and prevented enzymatic breakdown	[64]
Lidocaine docusate	Lidocaine docusate, an API in its IL form, exhibited greater solubility than the parent molecule	As an analgesic	Increased solubility, effectiveness, and thermal stability	[23]
1-Methyl-3-butyl-imidazolium ibuprofenate and lidocainium ibuprofenate	The release pattern of API-IL increased as a result of its integration into the polymer voids	To enhance drug delivery	Enhanced pattern of drug release	[65]

Table 1.
Applications of ionic liquids in drug delivery systems.

150°C for 2 minutes, the proposed reaction, which uses potassium hydroxide as the base and IL as solvent, was able to improve the typical reaction yield (70–91%) to 95%. This approach results in an aqueous potassium chloride solution as the only chemical waste produced during the process, and it is simple to extract the API product and recycle and reuse the solvent (**Figure 7a**) [67]. The hydrogenation of 2-arylacrilic acids in [C_4MIM] [BF_4] medium was published in 1997. The method to manufacture the nonsteroidal anti-inflammatory drug (NSAID) (S)-naproxen was used by the authors [69]. The electrosynthesis of naproxen by the electrocarboxylation of 2-(1-chloroethyl)-6-methoxynaphthalene using CO_2 employed [C_4C_1im][BF_4] as a reaction media (**Figure 7b**). When the solvent recovery was taken into account, this process produced high yields (89%) and conversion rates (90%), with 65% of atom economy. Even though the newly modified synthesis routes also enable the attainment of comparable high yields, the process in IL media employs CO_2 rather than CO, a well-known pollutant, and less expensive and more readily available catalysts (electron), which helps to develop "greener" routes for the synthesis of APIs [70]. Because of their special solvent qualities, ILs can act as both media and catalysts at the same time. For, example, a straightforward and effective iodination technique has been used to generate antifungal and antiprotozoal drugs, such as iodoquinol and clioquinol, respectively. In the absence of any oxidant, catalyst, or base, the IL ([C_4C_1py][DCI]) was utilized as both a solvent and an iodinating agent. By adding ICl (1.2 eq.), it is possible to regenerate the IL for up to five runs without losing its iodinating activity with a yield of >90% (**Figure 7c**) [71]. ILs can produce rapid microwave heating because of their charged composition, which leads to rapid and more efficient reactions. In a one-pot reaction, for instance, this method has been effectively used to directly lactamize lactones with >80% yields (**Figure 7d**) [72].

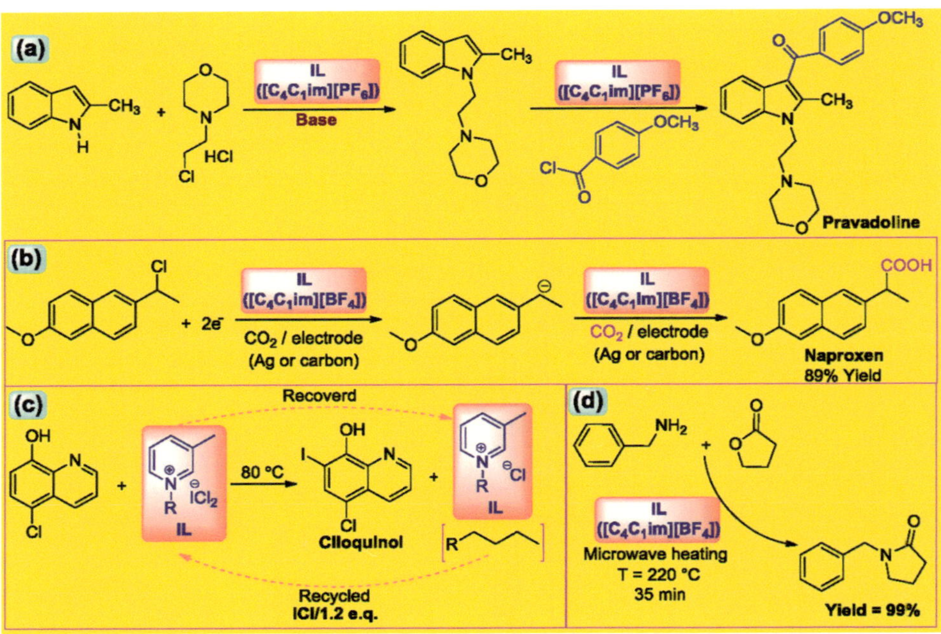

Figure 7.
(a) Synthesis of pravadoline in an IL media; (b) IL-based electrochemical production of naproxen; (c) synthesis of clioquinol in IL media; and (d) rapid and acid-free one-pot IL-based microwave methodology for direct synthesis of lactams.

Imidazolium salts are the most commonly utilized ionic liquids (ILs). **Table 2** shows that ILs are typically used as reaction media, which frequently functions as catalysts. Thus far, there have been descriptions of the synthesis of several NSAIDs [69, 71, 87], cholinesterase inhibitors [88], antiviral agents [73, 89], antimalarial [74], anticancer (**Figure 4a**) [89], antibacterial agents (**Figure 5**) [71, 90, 91], and radiolabeled molecular imaging and therapeutic agents [92]. Important examples of innovative medicinal compound synthesis using IL medium are displayed in **Table 2**.

3.2 Active Pharmaceutical Ingredient-Ionic Liquid (API-IL)

Active Pharmaceutical Ingredient-Ionic Liquids (API-ILs) are ionic liquids made from drug molecules, which are a novel approach to addressing the issues of bioavailability, low solubility, and thermal stability associated with traditional pharmaceutics. The idea of API-ILs is to preserve the drug's profile while giving it access to the advantageous characteristics of the solvent class and the counter-ion. The primary goal of the first wave of API-ILs was to synthesize these novel liquid salts and add well-known pharmacophores into them [23, 93]. Some examples of API-ILs are given in **Table 3**. Comparing IL-APIs to crystalline or solid pharmacological forms can reveal a number of benefits. One of the main issues in modern science is polymorphism, which can be resolved via IL-APIs [32]. Three methods exist for introducing API into IL systems (**Figure 8**). Type I involves utilizing API as an anion or cation through ionic binding; Type II requires covalent linkage; and Type III includes combining the two methods to create API-ILs with dual activity, allowing for the combination of comparable or different APIs in a single IL [33, 94]. The great majority of API-ILs that are currently on the market are Type I, meaning they have easily ionizable API moieties that can be utilized as IL anions or cations.

3.3 Application of ILs to drug extraction from natural products

Drug extraction from natural products uses a lot of volatile organic compounds (VOCs), which poses a major risk to the environment [95]. However, using ILs in place of conventional solvents may be able to reduce VOC emissions [96]. There have been numerous reports of recent attempts to use ILs as drug extraction solvents [97]. Protic ILs (PILs) were used to extract artemisinin, a strong naturally occurring antimalarial chemical, with water serving as the antisolvent [98]. Freire et al. have successfully carried out comprehensive extractions of prototypical alkaloids employing IL-based aqueous biphasic systems [97]. The findings indicate that the impact of impurity-causing agents (ILs) on the overall extraction of alkaloids is mostly determined by the characteristics of the cations rather than the anions. Materials that worked particularly well include [C_2OHMIM]Cl (increased hydrogen-bonding capabilities) and [C_7H_7MIM]Cl (improved aromatic interactions). The technique offers novel prospects for the concentration and isolation of other bioactive pharmaceuticals.

3.4 Toxicity and biocompatibility of ILs

The biodegradability and biocompatibility of ILs are generally unknown, which presents a challenge for their biological applications in pharmaceutics and the utilization of ILs as highly promising solvents for pharmaceutical and biological

IL	Role of IL	Compound	Activity	Ref.
[C_1OC_2MIM][Ms], [C_1OC_2 MIM] [TFa], [C_4MIM] [TFa]	Media	Stavudine	Anti-HIV drug	[73]
[C_1C_1MIM] [C_1SO_4]	Dehydrating agent	Hydrazinyl phthalazines	Anti-malarial agents	[74]
[C_1OC_2MIM][Ms], [C_1OC_2 MIM] [TFa], [C_4MIM] [TFa]	Media	Trifluridine	Anti-HSV drug	[73]
[C_2Py][BF_4] [MCl_m] (MCl_m = $AlCl_3$, $FeCl_3$, $ZnCl_2$, $SnCl_2$, $SnCl_4$, CuCl)	Catalyst	1-[*(E)*-3-methyl-4-benzenesulfonyl-3-methylbut-2-enyl]-2,3,4,5-tetramethoxy-6-methylbenzene	Coenzyme Q_{10} intermediate	[75]
[C_2MIM][Br]	Media for enzymatic enantioselective esterification	*(R)*-modafinil	Wakefulness-promoting agent	[76]
[C_4MIM][PF_6]	Media	Pravadoline	NSAID	[68]
[C_4C_1Py][ICl_2]	Iodinating reagent	Clioquinol	Antifungal drug	[71]
[C_4MIM][PF_6]	IL phase, whole-cell catalysis	*(R)*-1-trimethylsilylethanol	Vital synthon for different silicon-containing drugs	[77]
[C_4MIM][NTf_2]	IL phase, whole-cell catalysis	(S)-3-chloro-1-phenyl-1-propanol	Precursor of antidepressant drugs	[78]
[C_4C_1Py][ICl_2]	Iodinating reagent	Iodoquinol	Antiprotozoal drug	[71]
[C_4MIM][PF_6]	Media for enzymatic enantioselective esterification	11α-hydroxy-16α,17-epoxyprogesterone	Intermediate in the steroidal drug-synthesis process	[79]
[C_4MIM][BF_4]	Media	Isoxazolines	Precursor of antimicrobial agents	[80]
[C_4MIM][NTf_2]	Media	3-phenylglycidol	Precursor of various drugs, e.g., tomoxetine and reboxetine	[81]
[C_4MIM][PF_6], [C_8C_1Py][BF_4]	Media for enzymatic enantioselective esterification	*(R,S)*-1-chloro-3-(3,4-difluorophenoxy)-2-propanol	Intermediate in the synthesis of *(S)*-lubeluzole	[82]
[C_4MIM][PF_6]	IL phase, whole-cell catalysis	*(R)*-phenyl-acetylcarbinol	Precursor of (1*R*,2S)ephedrine and (1S,2S) pseudoephedrine	[83]
[C_4MIM][PF_6]	Media	Modafinil and its derivatives	Stimulant for treatment of sleep disorders	[84]

IL	Role of IL	Compound	Activity	Ref.
$[C_4MIM][BF_4]$	Media	*(S)*-naproxen	NSAID	[69]
$[C_5MIM][NO_3]$	Catalyst, media	α-tocopherol succinate	Vitamin E ester	[85]
Various imidazolium and pyridinium ILs	Media	Dronedarone intermediates	Anticardiac arrhythmia drug	[86]

Table 2.
Application of ILs in the synthesis of drugs and their intermediates.

IL-APIs		
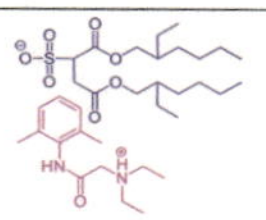 [Lidocainium][Docustae] Emollient, pain reliever	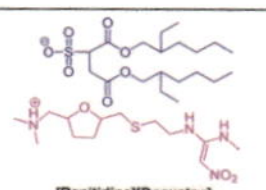 [Ranitidine][Docustae] Histamine H_2-receptor antagonist, emollient	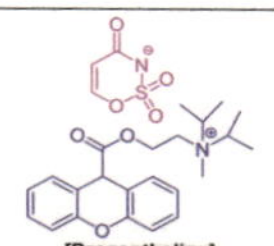 [Propantheline] [Acesulfamate] Antimuscarrinic, sweetener
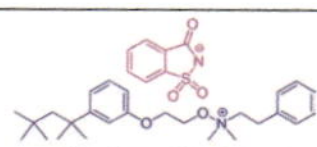 [Benzethonium][Saccharinate] Antiseptic, sweetener	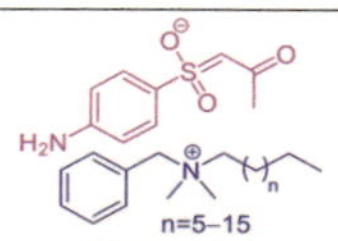 [Benzaalkoium] [Sulfacetamide] Antibacterial, anti-acne	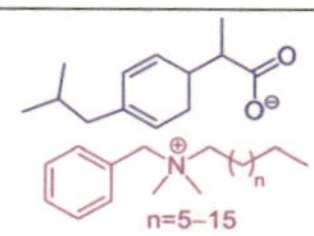 [Benzaalkoium][Ibuprofenate] Antibacterial, anti-inflammatory

Table 3.
Some examples of IL-APIs with their biological properties.

applications [99]. In order to be employed in biological applications, particularly for medicinal purposes, as solvents, co-solvents, or reagents, ILs should ideally have strong biodegradability and non-toxicity. Therefore, it is critical to evaluate the effects of ILs with respect to their environmental fate, ecotoxicity, bioaccumulation, and biodegradability [100–103].

To find non-toxic ILs, one approach is to modify the anion/cation combination of ILs, which can be utilized to fine-tune their chemical and biological properties. Fortunately, biocompatible organic cations and inorganic anions can be used to synthesize non-toxic ILs [104–106]. According to Amde and colleagues [103], the relative toxicity of the cationic head groups often follows the following trend:

Phosphonium > ammonium > imidazolium = pyridinium > morpholinium > pyrolidinium > piperidinium > choline.

There is still room for more research on the general toxicity, particularly cytotoxicity. Instead of restricting their usefulness, some ILs' high toxicity may make them promising antitumor and anticancer medicines, giving new opportunities in a variety of medical specialties.

3.5 Biodegradability and biocompatibility of ILs

The synthesis of biodegradable ILs has garnered more attention from researchers recently. The design of biocompatible ILs naturally coincides with the objective of biodegradability [107]. For instance, despite demonstrating low toxicity, ILs with a short alkyl chain attached to the cation have demonstrated resistance to

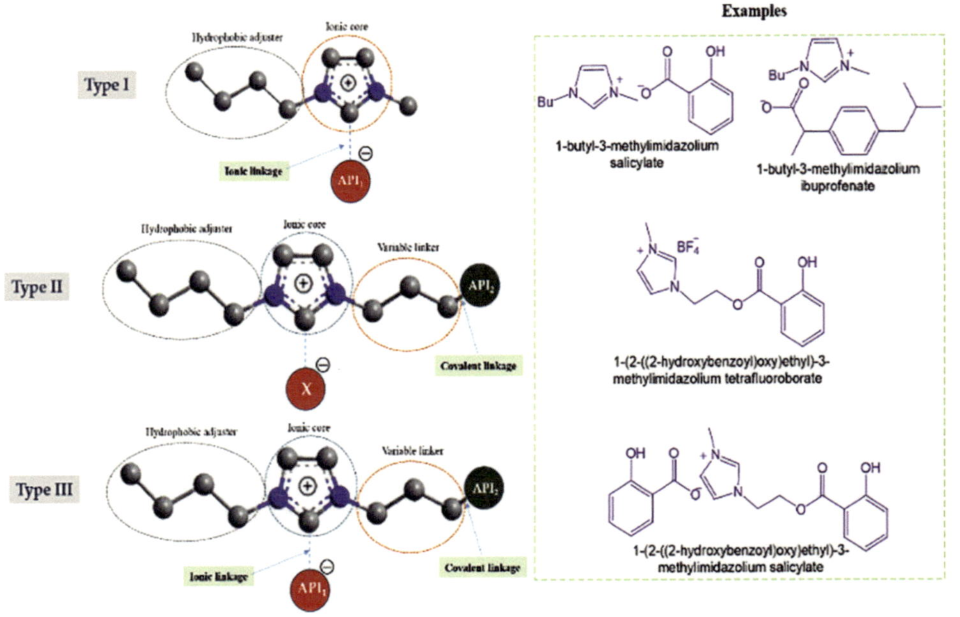

Figure 8.
Type I API-ILs contain ionic API as the anion; type II API-ILs comprise covalently coupled API as the cation; and type III API-ILs combine both possibilities. It should be noted that covalent binding to the anion is also plausible in type II (omitted for clarity) [8].

biodegradation [108]. It has been demonstrated that ILs with cationic groups (such phosphonium, pyridinium, ammonium, and imidazolium) that are especially toxic to microbes also degrade very quickly in the environment [109].

ILs with longer alkyl side chains (>C6) in the cationic core have been revealed to be biodegradable, while the same head group with shorter side chains is very weakly biodegradable. When it comes to their fundamental structure, imidazolium-based ILs generally show lower biodegradability, while 1-alkyl-3-methylpyridinium-based ILs have a higher potential for biodegradation [110].

On the other hand, the biodegradability of pyridinium ILs with linear alkyl chains is noticeably reduced [111]. The biodegradability of ILs is influenced not only by the cations but also by the nature of the anions. In contrast to ILs with shorter linear-chain anions (ethanoate and propanoate) those with longer linear-chain anions (butanoate, pentanoate, hexanoate, and octanoate) were found to be entirely biodegradable. On the other hand, it is known that anions like [NTf_2], [$N(CN)_2$], and [$B(CN)_4$] cannot be hydrolyzed or biodegraded in the environment.

4. Conclusions

The significance and benefits of ILs as feasible solvents in pharmaceutical applications have been emphasized in this chapter in comparison to traditional organic solvents. By taking advantage of their diverse ionic structures, basic characteristics of ILs can be altered and customized to meet desired requirements in biomedical applications. Developments in this area have shown that ILs can enhance the penetration of small and macromolecular pharmaceutical compounds through physiological barriers and solubilize water-insoluble or sparingly soluble drug molecules. In order to increase the stability of formulations and the transportation of drugs to the targeted site of action, ILs are showing as important components and drug binders. The solubility and polymorphism issues with solid drugs which were significant obstacles to the development of adaptive drug delivery methods have been effectively resolved by ILs with APIs. To address a number of problems and difficulties in the formulation and administration of IL-based therapeutic formulations, more research is necessary. We firmly believe that the tunability that is inherent in the realm of ILs is perfectly suitable and relevant to the pharmaceutical industry. In addition to offering remedies to issues that solid drugs frequently encounter, ILs can offer novel therapeutic or delivery possibilities that cannot be achieved by using solid APIs or conventional methods. A variety of IL-APIs that can be tailored to meet specific requirements while adjusting the chemical and physical characteristics of these ILs should be able to be prepared using such a modular design. Importantly, one may be able to solve the particular issue that caused this molecule to fail (such as poor solubility) by matching a failed API with a suitable cation or anion. Because of their inherent toxicity, lack of biodegradability, and detrimental effects on the environment, the therapeutic application of ILs is still unknown. Though the toxicity and biodegradability of ILs as green solvents remain unclear, a great deal of work has been made in the understanding and manufacture of biocompatible ILs for biological and pharmaceutical purposes. It has been discovered that a number of naturally occurring cations and anions that generate ILs, such as fatty acids, choline, and amino acids, are preferable to ammonium, imidazolium, and phosphonimum ILs. Further knowledge of ILs' structural characteristics and interactions with biosystems, particularly the coexistence of ILs and other living forms in an ecosystem, may also help to lessen their toxicity. In order to avoid having any effect on non-target areas or living forms, ILs must be designed with target specificity.

5. Future perspectives

Since structural features of ILs have been demonstrated to play a significant and often unexpected role in a variety of circumstances, developing pharmaceutical applications utilizing IL-based methodologies necessitates a thorough understanding of ILs both at the molecular and macroscopic levels. Furthermore, effective medications based on ILs might be given FDA approval in the near future. Even though ILs are no longer an innovative field, it is nevertheless remarkable how many different ways cations and anions can be combined to generate unique ILs with exciting new features. There has not been much work done to move the majority of IL examples from the laboratory to the bench scale; they are still proofs of concept. While the traditional pharmaceutical industry relies heavily on solids and solids processing, the use of ILs should be promoted as a way to recycle many of the drugs that have been shelved due to limited aqueous solubility or polymorphic conversion, giving these products a new market value. The pharmaceutical sector can undergo unexpected changes if a modular IL approach is implemented. This strategy can offer a foundation for enhanced performance with novel therapeutic approaches or even individualized drug regimens. New and fascinating findings will continue to be produced by the knowledge foundation created by international efforts in IL research as well as the research community's openness to explore new things. We propose that pharmaceutical companies should follow the increasing number of industries that are thinking about using ILs for more than simply solvents in their processes. We think that in the near future, clinical trials involving green and biocompatible ILs to advance IL-based medication delivery technology will be carried out, which will stimulate the usage of ILs in commercial applications.

Conflict of interest

The authors declare no conflict of interest.

List of abbreviations

ILs	ionic liquid
DDSs	drug delivery systems
PILs	protic ionic liquid
TSILs	task-specific ILs
HIV	human immunodeficiency virus
[C_1SO_4]	methyl sulfate
[C2Py]	1-ethylpyridinium
[Lac]	lactate
[BF_4]	tetrafluoroborate
[PF_6]	hexafluorophosphate
[Ms]	methanesulfonate
[TFa]	trifluoroacetate
[Pro]	propanoate
[Br]	bromide
[ICl_2]	dichloroiodate
[$(C_2)_3NH$]	triethylammonium
RTILs	room-temperature ILs
[OAc]	acetate
[NO_3]	nitrate
PIL	polyionic liquids
ST	surface tension
CAGE	choline-geranate
[IBut]	isobutanoate
[C_1C_1MIM]	1,2,3-trimethylimidazolium
APIs	active pharmaceutical ingredients
VOCs	volatile organic compounds
SILMs	supported IL membranes
MOFs	metal-organic frameworks
HSV	herpes simplex virus
NSAID	nonsteroidal anti-inflammatory drug
[C_2MIM]	1-ethyl-3-methylimidazolium
[C_1OC_2MIM]	1-methoxyethyl-3-methylimidazolium
[C_7H_7MIM]	1-benzyl-3-methylimidazolium
[C_2OHMIM]	1-hydroxyethyl-3-methylimidazolium
[C_5MIM]	1-pentyl-3-methylimidazolium
[$(C_1)_2(C_1)_2G$]	1,1,3,3-tetramethylguanidine

[Cl]	chloride
[C_4C_1Py]	1-butyl-3-methylpyridinium
[NTf_2]	bis(trifluoromethylsulfonyl)amide
[HDBU]	1,8-diazabicyclo[5.4.0]undec-7-en-8-ium
[$N(CN)_2$]	dicyanamide
[HSO_4]	hydrogensulfate
[C_8C_1Py]	1-octyl-3-methylpyridinium
TFA	trifluroacetic acid
TFSA	trifluoromethanesulfonic acid
[C_4MIM]	1-butyl-3-methylimidazolium
([C4C1py][DCI])	1-butyl-3-methylpyridinium dichloroiodate

Author details

Sadia Rani[1], Ali Irfan[2*], Samina Aslam[1*], Gamal A. Shazly[3], Huma Hameed[4], Muhammad Zaman[4], Noreen Samad[5] and Yousef A. Bin Jardan[3*]

1 Department of Chemistry, The Women University Multan, Pakistan

2 Department of Chemistry, Government College University Faisalabad, Faisalabad, Pakistan

3 Department of Pharmaceutics, College of Pharmacy, King Saud University, Riyadh, Saudi Arabia

4 Faculty of Pharmaceutical Sciences, University of Central Punjab, Lahore, Pakistan

5 Department of Biochemistry, Bahauddin Zakariya University, Multan, Pakistan

*Address all correspondence to: raialiirfan@gmail.com, drsamina.chem@wum.edu.pk and ybinjardan@ksu.edu.sa

References

[1] Moshikur RM, Chowdhury MR, Moniruzzaman M, Goto M. Biocompatible ionic liquids and their application in pharmaceutics. Green Chemistry. 2020;**22**:8116-8139. DOI: 10.1039/D0GC02387F

[2] Stella VJ. Prodrugs: Some thoughts and current issues. Journal of Pharmaceutical Sciences. 2010;**99**:4755-4765. DOI: 10.1002/jps.22205

[3] Huttunen KM, Raunio H, Rautio J. Prodrugs—From serendipity to rational design. Pharmacological Reviews. 2011;**63**:750-771. DOI: 10.1124/pr.110.003459

[4] Torchilin VP. Recent advances with liposomes as pharmaceutical carriers. Nature Reviews. Drug Discovery. 2005;**4**:145-160. DOI: 10.1038/nrd1632

[5] Serajuddin ATM. Salt formation to improve drug solubility. Advanced Drug Delivery Reviews. 2007;**59**:603-616. DOI: 10.1016/j.addr.2007.05.010

[6] Vasconcelos T, Sarmento B, Costa P. Solid dispersions as strategy to improve oral bioavailability of poor water soluble drugs. Drug Discovery Today. 2007;**12**:1068-1075. DOI: 10.1016/j.drudis.2007.09.005

[7] Dean PM, Turanjanin J, Yoshizawa-Fujita M, MacFarlane DR, Scott JL. Exploring an anti-crystal engineering approach to the preparation of pharmaceutically active e ionic liquids. Crystal Growth & Design. 2009;**9**:1137-1145. DOI: 10.1021/cg8009496

[8] Egorova KS, Gordeev EG, Ananikov VP. Ananikov biological activity of ionic liquids and their application in pharmaceutics and medicine. Chemical Reviews. 2017;**117**:7132-7189. DOI: 10.1021/acs.chemrev.6b00562

[9] Procopio D, Siciliano C, Trombino S, Dumitrescu DE, Suciu F, Di Gioia ML. Green solvents for the formation of amide linkages. Organic & Biomolecular Chemistry. 2022;**20**:1137-1149. DOI: 10.1039/D1OB01814K

[10] Curreri AM, Mitragotri S, Tanner EE. Recent advances in ionic liquids in biomedicine. Advancement of Science. 2021;**8**:2004819. DOI: 10.1002/advs.202004819

[11] Wilkes JS, Zaworotko MJJ. Air and water stable 1-ethyl-3-methylimidazolium based ionic liquids. Journal of the Chemical Society, Chemical Communications. 1992:965-967. DOI: 10.1039/C39920000965

[12] Curreri AM, Mitragotri S. Recent advances in ionic liquids in biomedicine. Advancement of Science. 2021;**8**:2004819. DOI: 10.1002/advs.202004819

[13] Md Moshikur R, Goto M. Pharmaceutical applications of ionic liquids: A personal account. Chemical Record. 2023;**23**:e202300026. DOI: 10.1002/tcr.202300026

[14] Warner IM, El-Zahab B, Siraj N. Perspectives on moving ionic liquid chemistry into the solid phase. Analytical Chemistry. 2014;**86**:7184-7191. DOI: 10.1021/ac501529m

[15] Ravula S, Wise KW, Shinde PS, Bara JE. Design and performance of di- and tricationic poly (ionic liquid) + ionic liquid composite membranes for CO_2 separation. Macromolecules.

2023;**56**:6126-6141. DOI: 10.1021/acs.macromol.3c00850

[16] Li X, Chen K, Guo R, Wei Z. Ionic liquids functionalized MOFs for adsorption. Chemical Reviews. 2023;**123**:10432-10467. DOI: 10.1021/acs.chemrev.3c00248

[17] Li G, Chen K, Lei Z, Wei Z. Condensable gases capture with ionic liquids. Chemical Reviews. 2023;**123**:10258-10301. DOI: 10.1021/acs.chemrev.3c00175

[18] Black IM, Ndukwe IE, Vlach J, Backe J, Urbanowicz BR, Heiss C, et al. Acetylation in ionic liquids dramatically increases yield in the glycosyl composition and linkage analysis of insoluble and acidic polysaccharides. Analytical Chemistry. 2023;**95**:12851-12858. DOI: 10.1021/acs.analchem.3c02056

[19] Xie T, Lv X, Tian S, Xie Y, Lv A, Lv Z, et al. Simple fabrication of silica amino sphere-reinforced ionic liquids/graphene conductive hydrogel sensors with super toughness self-healing and strain sensitivity properties. Macromolecules. 2023;**56**:6256-6266. DOI: 10.1021/acs.macromol.3c00496

[20] Zhao H. Current studies on some physical properties of ionic liquids. Physics and Chemistry of Liquids. 2003;**41**:545-557. DOI: 10.1080/003191031000117319

[21] Huang W, Wu X, Qi J, Zhu Q, Wu W, Lu Y, et al. Ionic liquids: Green and tailor-made solvents in drug delivery. Drug Discovery Today. 2020;**25**:901-908. DOI: 10.1016/j.drudis.2019.09.018

[22] Dinis TB, Silva FA, Sousa F, Freire MG. Advances brought by hydrophilic ionic liquids in fields involving pharmaceuticals. Materials. 2021;**14**:6231. DOI: 10.3390/ma14216231

[23] Hough WL, Smiglak M, Rodríguez H, Swatloski RP, Spear SK, Daly DT, et al. The third evolution of ionic liquids: Active pharmaceutical ingredients. New Journal of Chemistry. 2007;**31**:1429-1436. DOI: 10.1039/B706677P

[24] Caparica R, Júlio A, Rosado C, Santos de Almeida T. Applicability of ionic liquids in topical drug delivery systems: A mini review. Journal of Pharmacology and Clinical Research. 2018;**4**:555649. DOI: 10.19080/JPCR.2018.04.555649

[25] MacFarlane DR, Chong AL, Forsyth M, Kar M, Vijayaraghavan R, Somers A, et al. New dimensions in salt–solvent mixtures: A 4th evolution of ionic liquids. Faraday Discussions. 2018;**206**:9-28. DOI: 10.1039/C7FD00189D

[26] Dias AR, Costa-Rodrigues J, Fernandes MH, Ferraz R, Prudêncio C. The anticancer potential of ionic liquids. ChemMedChem. 2017;**12**:11-18. DOI: 10.1002/cmdc.201600480

[27] Sheldon RA. Biocatalysis and biomass conversion in alternative reaction media. Chemistry - A European Journal. 2016;**22**:12984-12999. DOI: 10.1002/chem.201601940

[28] Welton T. Ionic liquids: A brief history. Biophysical Reviews. 2018;**10**:691-706. DOI: 10.1007/s12551-018-0419-2

[29] Ferraz R, Branco LC, Prudencio C, Noronha JP, Petrovski Z. Ionic liquids as active pharmaceutical ingredients. ChemMedChem. 2011;**6**:975-985. DOI: 10.1002/cmdc.201100082

[30] Dias AR, Costa-Rodrigues J, Teixeira C, Prudêncio C, Gomes P, Ferraz R. Ionic liquids for topical delivery

in cancer. Current Medicinal Chemistry. 2019;**26**:7520-7532. DOI: 10.2174/0929867325666181026110227

[31] Elgharbawy AA, Moniruzzaman M, Goto M. Recent advances of enzymatic reactions in ionic liquids: Part II. Biochemical Engineering Journal. 2020;**154**:107426. DOI: 10.1016/j.bej.2019.107426

[32] Wu X, Zhu Q, Chen Z, Wu W, Lu Y, Qi J. Ionic liquids as a useful tool for tailoring active pharmaceutical ingredients. Journal of Controlled Release. 2021;**338**:268-283. DOI: 10.1016/j.jconrel.2021.08.032

[33] Shamshina JL, Barber PS, Rogers RD. Ionic liquids in drug delivery. Expert Opinion on Drug Delivery. 2013;**10**:1367-1381. DOI: 10.1517/17425247.2013.808185

[34] Agatemor C, Ibsen KN, Tanner EE, Mitragotri S. Bioengineering & Translational Medicine, Ionic liquids for addressing unmet needs in healthcare. 2018;**3**:7-25. DOI: 10.1002/btm2.10083

[35] Egorova KS, Ananikov VP. Fundamental importance of ionic interactions in the liquid phase: A review of recent studies of ionic liquids in biomedical and pharmaceutical applications. Journal of Molecular Liquids. 2018;**272**:271-300. DOI: 10.1016/j.molliq.2018.09.025

[36] Moshikur RM, Chowdhury MR, Wakabayashi R, Tahara Y, Moniruzzaman M, Goto M. Ionic liquids with methotrexate moieties as a potential anticancer prodrug: Synthesis characterization and solubility evaluation. Journal of Molecular Liquids. 2019;**278**:226-233. DOI: 10.1016/j.molliq.2019.01.063

[37] Heckenbach ME, Romero FN, Green MD, Halden RU. Meta-analysis of ionic liquid literature and toxicology. Chemosphere. 2016;**150**:266-274. DOI: 10.1016/j.chemosphere.2016.02.029

[38] Krossing I, Slattery JM, Daguenet C, Dyson PJ, Oleinikova A, Weingärtner H. Why are ionic liquids liquid? A simple explanation based on lattice and solvation energies. Journal of the American Chemical Society. 2006;**128**:13427-13434. DOI: 10.1021/ja0619612

[39] Gandhewar N, Shende P. Ionic liquids: A state of the art for biomedical applications. Ionics. 2021;**27**:3715-3728. DOI: 10.1007/s11581-021-04201-y

[40] Baker SN, McCleskey TM, Pandey S, Baker GA. Fluorescence studies of protein thermostability in ionic liquids. Chemical Communications. 2004:940-941. DOI: 10.1039/B401304M

[41] Erbeldinger M, Mesiano AJ, Russell A. Enzymatic catalysis of formation of Z-aspartame in ionic liquid– an alternative to enzymatic catalysis in organic solvents. Journal of Biotechnology Progress. 2000;**16**:1129-1131. DOI: 10.1021/bp000094g

[42] Moniruzzaman M, Tahara Y, Tamura M, Kamiya N, Goto M. Ionic liquid-assisted transdermal delivery of sparingly soluble drugs. Chemical Communications. 2010;**46**:1452-1454. DOI: 10.1039/B907462G

[43] Jaitely V, Karatas A, Florence AT. Water-immiscible room temperature ionic liquids (RTILs) as drug reservoirs for controlled release. International Journal of Pharmaceutics. 2008;**354**:168-173. DOI: 10.1016/j.ijpharm.2008.01.034

[44] Ali MK, Moshikur RM, Goto M, Moniruzzaman M. Recent developments in ionic liquid-assisted topical and transdermal drug delivery.

Pharmaceutical Research. 2022;**39**:2335-2351. DOI: 10.1007/s11095-022-03322-x

[45] Qader IB, Prasad K. Recent developments on ionic liquids and deep eutectic solvents for drug delivery applications. Pharmaceutical Research. 2022;**39**:2367-2377. DOI: 10.1007/s11095-022-03315-w

[46] Jiang L, Sun Y, Lu A, Wang X, Shi Y. Ionic liquids: Promising approach for oral drug delivery. Pharmaceutical Research. 2022;**39**:2353-2365. DOI: 10.1007/s11095-022-03260-8

[47] Gomes A, Aguiar L, Ferraz R, Teixeira C, Gomes P. The emerging role of ionic liquid-based approaches for enhanced skin permeation of bioactive molecules: A snapshot of the past couple of years. International Journal of Molecular Sciences. 2021;**22**:11991. DOI: 10.3390/ijms222111991

[48] Nikfarjam N, Ghomi M, Agarwal T, Hassanpour M, Sharifi E, Khorsandi D, et al. Antimicrobial ionic liquid-based materials for biomedical applications. Advanced Functional Materials. 2021;**31**:2104148. DOI: 10.1002/adfm.202104148

[49] Pernak J, Sobaszkiewicz K, Mirska I. Anti-microbial activities of ionic liquids. Green Chemistry. 2003;**5**:52-56. DOI: 10.1039/B207543C

[50] Gomes JM, Silva SS, Reis RL. Biocompatible ionic liquids: Fundamental behaviours and applications. Chemical Society Reviews. 2019;**48**:4317-4335. DOI: 10.1039/D2CS00976E

[51] Kumar V, Malhotra SV. Antitumor activity of ionic liquids on human tumor cell lines. In: Malhotra SV, editor. Ionic Liquid Applications: Pharmaceuticals, Therapeutics, and Biotechnology. ACS Symposium Series: ACS Publications; 2010. pp. 91-102. DOI: 10.1021/bk-2010-1038.ch008.ch8

[52] Kaushik NK, Attri P, Kaushik N, Choi EH. Synthesis and antiproliferative activity of ammonium and imidazolium ionic liquids against T98G brain cancer cells. Molecules. 2012;**17**:13727-13739. DOI: 10.3390/molecules171213727

[53] Busetti A, Crawford DE, Earle MJ, Gilea MA, Gilmore BF, Laverty GSP, et al. Antimicrobial and antibiofilm activities of 1-alkylquinolinium bromide ionic liquids. Green Chemistry. 2010;**12**:420-425. DOI: 10.1039/B919872E

[54] Banerjee A, Ibsen K, Brown T, Chen R, Agatemor C, Mitragotri S. Ionic liquids for oral insulin delivery. Proceedings of the National Academy of Sciences. 2018;**115**:7296-7301. DOI: 10.1073/pnas.1722338115

[55] Huang W, Fang Z, Zheng X, Qi J, Wu W, Lu Y. Green and controllable fabrication of nanocrystals from ionic liquids. Chinese Chemical Letters. 2022;**33**:4079-4083. DOI: 10.1016/j.cclet.2022.01.043

[56] Pang J, Luan Y, Li F, Cai X, Li Z. Ionic liquid-assisted synthesis of silica particles and their application in drug release. Materials Letters. 2010;**64**:2509-2512. DOI: 10.1016/j.matlet.2010.08.032

[57] Slowing II, Vivero-Escoto JL, Wu C-W, Lin VSY. Mesoporous silica nanoparticles as controlled release drug delivery and gene transfection carriers. Advanced Drug Delivery Reviews. 2008;**60**:1278-1288. DOI: 10.1016/j.addr.2008.03.012

[58] Mahajan S, Sharma R, Mahajan RK. An investigation of drug binding ability of a surface active ionic liquid:

Micellization electrochemical and spectroscopic studies. Langmuir. 2012;**28**:17238-17246. DOI: 10.1021/la303193n

[59] Cui W, Lu X, Cui K, Niu L, Wei Y, Lu Q. Dual-responsive controlled drug delivery based on ionically assembled nanoparticles. Langmuir. 2012;**28**:9413-9420. DOI: 10.1021/la3016436

[60] Zhou G, Luo Z, Fu X. Preparation of starch nanoparticles in a water-in-ionic liquid microemulsion system and their drug loading and releasing properties. Journal of Agricultural and Food Chemistry. 2014;**62**:8214-8220. DOI: 10.1021/jf5018725

[61] de Almeida TS, Júlio A, Mota JP, Rijo P, Reis CP. An emerging integration between ionic liquids and nanotechnology: General uses and future prospects in drug delivery. Therapeutic Delivery. 2017;**8**:461-473. DOI: 10.4155/tde-2017-0002

[62] Caparica R, Júlio A, Araújo MEM, Baby AR, Fonte P, Costa JG, et al. Anticancer activity of rutin and its combination with ionic liquids on renal cells. Biomolecules. 2020;**10**:233. DOI: 10.3390/biom10020233

[63] Singla P, Singh O, Chabba S, Mahajan RK. Pluronic-SAILs (surface active ionic liquids) mixed micelles as efficient hydrophobic quercetin drug carriers. Journal of Molecular Liquids. 2018;**249**:294-303. DOI: 10.1016/j.molliq.2017.11.044

[64] Hattori T, Tagawa H, Inai M, Kan T, Si K, Itai S, et al. Transdermal delivery of nobiletin using ionic liquids. Scientific Reports. 2019;**9**:20191. DOI: 10.1038/s41598-019-56731-1

[65] Jouannin C, Tourné-Péteilh C, Darcos V, Sharkawi T, Devoisselle JM, Gaveau P, et al. Drug delivery systems based on pharmaceutically active ionic liquids and biocompatible poly (lactic acid). Journal of Materials Chemistry B. 2014;**2**:3133-3141. DOI: 10.1039/C4TB00264D

[66] Martins MA, Frizzo CP, Moreira DN, Zanatta N, Bonacorso HG. Ionic liquids in heterocyclic synthesis. Chemical Reviews. 2008;**108**:2015-2050. DOI: 10.1021/cr078399y

[67] Pedro SN, Freire CSR, Silvestre AJD, Freire MG. The role of ionic liquids in the pharmaceutical field: An overview of relevant applications. International Journal of Molecular Sciences. 2020;**21**:8298. DOI: 10.3390/ijms21218298

[68] Earle MJ, McCormac PB, Seddon KR. The first high yield green route to a pharmaceutical in a room temperature ionic liquid. Green Chemistry. 2000;**2**:261-262. DOI: 10.1039/B006612P

[69] Monteiro AL, Zinn FK, de Souza RF, Dupont. Asymmetric hydrogenation of 2-arylacrylic acids catalyzed by immobilized Ru-BINAP complex in 1-n-butyl-3-methylimidazolium tetrafluoroborate molten salt. Tetrahedron: Asymmetry. 1997;**8**:177-179. DOI: 10.1016/S0957-4166(96)00485-5

[70] Mena S, Santiago S, Gallardo I, Guirado G. Sustainable and efficient electrosynthesis of naproxen using carbon dioxide and ionic liquids. Chemosphere. 2020;**245**:125557. DOI: 10.1016/j.chemosphere.2019.125557

[71] Deshmukh A, Gore B, Thulasiram HV, Swamy VP. Recyclable ionic liquid iodinating reagent for solvent free regioselective iodination of activated aromatic and heteroaromatic amines. RSC Advances. 2015;**5**:88311-88315. DOI: 10.1039/C5RA14702F

[72] Orrling KM, Wu X, Russo F, Larhed M. Fast acid-free and selective lactamization of lactones in ionic liquids. The Journal of Organic Chemistry. 2008;**73**:8627-8630. DOI: 10.1021/jo8015264

[73] Kumar V, Malhotra SV. Synthesis of nucleoside-based antiviral drugs in ionic liquids. Bioorganic & Medicinal Chemistry Letters. 2008;**18**:5640-5642. DOI: 10.1016/j.bmcl.2008.08.090

[74] Subramanian G, Rajeev CB, Mohan CD, Sinha A, Chu TT, Anusha S, et al. Synthesis and in vitro evaluation of hydrazinyl phthalazines against malaria parasite *Plasmodium falciparum*. Bioorganic & Medicinal Chemistry Letters. 2016;**26**:3300-3306. DOI: 10.1016/j.bmcl.2016.05.049

[75] Chen Y, Zu Y, Fu Y, Zhang X, Yu P, Sun G, et al. Efficient Lewis acid ionic liquid-catalyzed synthesis of the key intermediate of coenzyme Q10 under microwave irradiation. Molecules. 2010;**15**:9486-9495. DOI: 10.3390/molecules15129486

[76] Gao F, Wang L, Liu Y, Wang S, Jiang Y, Hu M, et al. Enzymatic synthesis of (R)-modafinil by chloroperoxidase-catalyzed enantioselective sulfoxidation of 2-(diphenylmethylthio) acetamide. Biochemical Engineering Journal. 2015;**93**:243-249. DOI: 10.1016/j.bej.2014.10.017

[77] Zhang BB, Cheng J, Lou WY, Wang P, Zong MH. Efficient anti-Prelog enantioselective reduction of acetyltrimethylsilane to (R)-1-trimethylsilylethanol by immobilized *Candida parapsilosis* CCTCC M203011 cells in ionic liquid-based biphasic systems. Microbial Cell Factories. 2012;**11**:1-12. DOI: 10.1186/1475-2859-11-108

[78] Choi HJ, Uhm KN, Kim HK. Production of chiral compound using recombinant *Escherichia coli* cells co-expressing reductase and glucose dehydrogenase in an ionic liquid/water two phase system. Journal of Molecular Catalysis B: Enzymatic. 2011;**70**:114-118. DOI: 10.1016/j.molcatb.2011.02.013

[79] Wu DX, Guan YX, Wang HQ, Yao SJ. 11α-hydroxylation of 16α 17-epoxyprogesterone by Rhizopus nigricans in a biphasic ionic liquid aqueous system. Bioresource Technology. 2011;**102**:9368-9373. DOI: 10.1016/j.biortech.2011.07.060

[80] Chakraborty B, Sharma CD. A new route to the synthesis of isoxazoline derivatives from dihydropyran via cycloaddition reaction in ionic liquid. Tetrahedron Letters. 2013;**54**:5532-5536. DOI: 10.1016/j.tetlet.2013.07.159

[81] Kazemi S, Martín AB, Sordi D, Kroon MC, Peters CJ, Arends IW. Arends vanadium-catalyzed epoxidation reaction of cinnamyl alcohol in ionic liquids. Green Processing and Synthesis. 2012;**1**:509-516. DOI: 10.1515/gps-2012-0063

[82] Singh M, Singh RS, Banerjee UC. Stereoselective synthesis of (R)-1-chloro-3 (3 4-difluorophenoxy)-2-propanol using lipases from *Pseudomonas aeruginosa* in ionic liquid-containing system. Journal of Molecular Catalysis B: Enzymatic. 2009;**56**:294-299. DOI: 10.1016/j.molcatb.2008.06.002

[83] Kandar S, Suresh A, Noronha SB. (R)-PAC biosynthesis in [BMIM][PF_6]/aqueous biphasic system using *Saccharomyces cerevisiae* BY4741 cells. Applied Biochemistry and Biotechnology. 2015;**175**:1771-1788. DOI: 10.1007/s12010-014-1394-0

[84] Ternois J, Guillen F, Plaquevent JC, Coquerel G. Asymmetric synthesis

of modafinil and its derivatives by enantioselective oxidation of thioethers: Comparison of various methods including synthesis in ionic liquids. Tetrahedron: Asymmetry. 2007;**18**:2959-2964. DOI: 10.1016/j.tetasy.2007.11.031

[85] Tao Y, Dong R, Pavlidis IV, Chen B, Tan T. Using imidazolium-based ionic liquids as dual solvent-catalysts for sustainable synthesis of vitamin esters: Inspiration from bio-and organo-catalysis. Green Chemistry. 2016;**18**:1240-1248. DOI: 10.1039/c5gc02557e

[86] Kretzschmar G, Kraft V, Rossen K, Graeser J. Process for the production of dronedarone intermediates. Google Patents. 2014

[87] Hongwei Y, Jinchuan W, Chi BC. Kinetic resolution of ibuprofen catalyzed by *Candida rugosa* lipase in ionic liquids. Chirality. 2005;**17**:16-21. DOI: 10.1002/chir.20078

[88] Basiri A, Murugaiyah V, Osman H, Kumar RS, Kia Y, Hooda A, et al. Cholinesterase inhibitory activity versus aromatic core multiplicity: A facile green synthesis and molecular docking study of novel piperidone embedded thiazolopyrimidines. Bioorganic & Medicinal Chemistry. 2014;**22**:906-916. DOI: 10.1016/j.bmc.2013.11.020

[89] Liu BK, Wang N, Chen ZC, Wu Q, Lin XF. Markedly enhancing lipase-catalyzed synthesis of nucleoside drugs' ester by using a mixture system containing organic solvents and ionic liquid. Bioorganic & Medicinal Chemistry Letters. 2006;**16**:3769-3771. DOI: 10.1016/j.bmcl.2006.04.054

[90] Naik PU, Harjani JR, Nara SJ, Salunkhe MM. Ionic liquid enabled sulfamoylation of arenes: An ambient expeditious and regioselective protocol for aryl sulfonamides. Tetrahedron Letters. 2004;**45**:1933-1936. DOI: 10.1016/j.tetlet.2003.12.132

[91] Subhedar DD, Shaikh MH, Nawale L, Yeware A, Sarkar D, Shingate BB. [Et_3NH][HSO_4] catalyzed efficient synthesis of 5-arylidene-rhodanine conjugates and their antitubercular activity. Research on Chemical Intermediates. 2016;**42**:6607-6626. DOI: 10.1007/s11164-016-2484-0

[92] Rajerison H, Faye D, Roumesy A, Louaisil N, Boeda F, Faivre-Chauvet A, et al. Ionic liquid supported organotin reagents to prepare molecular imaging and therapy agents. Organic & Biomolecular Chemistry. 2016;**14**:2121-2126. DOI: 10.1039/C5OB02459E

[93] Bica K, Rijksen C, Nieuwenhuyzen M, Rogers RD. In search of pure liquid salt forms of aspirin: Ionic liquid approaches with acetylsalicylic acid and salicylic acid. Physical Chemistry Chemical Physics. 2010;**12**:2011-2017. DOI: 10.1039/B923855G

[94] Egorova KS, Seitkalieva MM, Posvyatenko AV, Khrustalev VN, Ananikov VP. Cytotoxic activity of salicylic acid-containing drug models with ionic and covalent binding. ACS Medicinal Chemistry Letters. 2015;**6**:1099-1104. DOI: 10.1021/acsmedchemlett.5b00258

[95] Bassoli A, Brambilla A, Bolzacchini E, Chioccara F, Morazzoni F, Orlandi M, et al. In: Anastas PT, Williamson TC, editors. Green Chemistry: Designing Chemistry for the Environment. ACS Symposium Series: ACS Publications; 1996. pp. 92-104. DOI: 10.1021/bk-1996-0626.ch8

[96] Dupont J, de Souza RF, Suarez PA. Liquid (molten salt) phase organometallic catalysis. Chemical Reviews. 2002;**102**:3667-3692. DOI: 10.1021/cr010338r

[97] Freire MG, Neves CM, Marrucho IM, Lopes JNC, Rebelo LPN, Coutinho JA. Coutinho high-performance extraction of alkaloids using aqueous two-phase systems with ionic liquids. Green Chemistry. 2010;**12**:1715-1718. DOI: 10.1039/C0GC00179A

[98] Sanders MW, Wright L, Tate L, Fairless G, Crowhurst L, Bruce NC, et al. Unexpected preferential dehydration of artemisinin in ionic liquids. The Journal of Physical Chemistry. A. 2009;**113**:10143-10145. DOI: 10.1021/jp906436e

[99] Moniruzzaman M, Kamiya N, Goto M. Ionic liquid based microemulsion with pharmaceutically accepted components: Formulation and potential applications. Journal of Colloid and Interface Science. 2010;**352**:136-142. DOI: 10.1016/j.jcis.2010.08.035

[100] Zhao H, Baker GA, Holmes S. New eutectic ionic liquids for lipase activation and enzymatic preparation of biodiesel. Organic & Biomolecular Chemistry. 2011;**9**:1908-1916. DOI: 10.1039/C0OB01011A

[101] Ventura SPM, Santos-Ebinuma VC, Pereira JFBP. Isolation of natural red colorants from fermented broth using ionic liquid-based aqueous two-phase systems. Journal of Industrial Microbiology & Biotechnology. 2013;**40**:507-516. DOI: 10.1007/s10295-013-1237-y

[102] Ventura SP, Gonçalves AM, Sintra T, Pereira JL, Gonçalves F, Coutinho JA. Designing ionic liquids: The chemical structure role in the toxicity. Ecotoxicology. 2013;**22**:1-12. DOI: 10.1007/s10646-012-0997-x

[103] Amde M, Liu JF, Pang L. Environmental application fate effects and concerns of ionic liquids: A review. Environmental Science & Technology. 2015;**49**:12611-12627. DOI: 10.1021/acs.est.5b03123

[104] Frade RF, Rosatella AA, Marques CS, Branco LC, Kulkarni PS, Mateus NM, et al. Toxicological evaluation on human colon carcinoma cell line (CaCo-2) of ionic liquids based on imidazolium guanidinium ammonium phosphonium pyridinium and pyrrolidinium cations. Green Chemistry. 2009;**11**:1660-1665. DOI: 10.1039/B914284N

[105] Radošević K, Bubalo MC, Srček VG, Grgas D, Dragičević TL, Redovniković IR. Evaluation of toxicity and biodegradability of choline chloride based deep eutectic solvents. Ecotoxicology and Environmental Safety. 2015;**112**:46-53. DOI: 10.1016/j.ecoenv.2014.09.034

[106] Pernak J, Borucka N, Walkiewicz F, Markiewicz B, Fochtman P, Stolte S, et al. Synthesis toxicity biodegradability and physicochemical properties of 4-benzyl-4-methylmorpholinium-based ionic liquids. Green Chemistry. 2011;**13**:2901-2910. DOI: 10.1039/C1GC15468K

[107] Moniruzzaman M, Goto MJ. Ionic liquids: Future solvents and reagents for pharmaceuticals. Journal of Chemical Engineering of Japan. 2011;**44**:370-381. DOI: 10.1252/jcej.11we015

[108] Gathergood N, Garcia MT, Scammells PJ. Biodegradable ionic liquids: Part I. Concept preliminary targets and evaluation. Green Chemistry. 2004;**6**:166-175. DOI: 10.1039/B315270G

[109] Wells AS, Coombe VT. On the freshwater ecotoxicity and biodegradation properties of some common ionic liquids. Organic Process Research and Development. 2006;**10**:794-798. DOI: 10.1021/op060048i

[110] Harjani JR, Singer RD, Garcia MT, Scammells PG. Scammells biodegradable pyridinium ionic liquids: Design synthesis and evaluation. Green Chemistry. 2009;**11**:83-90. DOI: 10.1039/b811814k

[111] Neumann J, Steudte S, Cho CW, Thöming J, Stolte S. Biodegradability of 27 pyrrolidinium morpholinium piperidinium imidazolium and pyridinium ionic liquid cations under aerobic conditions. Green Chemistry. 2014;**16**:2174-2184. DOI: 10.1039/C3GC41997E

Chapter 4

Synthesis and Ionic Conductivity of Dicationic Ionic Salts as Electrolytes for the Next Generation of Batteries Including Solid State Batteries

Pradip K. Bhowmik, Seonghyeok L. Cox,
Si L. Chen, Jakob Smith, David King, Haesook Han,
Thamil Selvi Velayutham and Alfonso Martinez-Felipe

Abstract

The insatiable demand for clean, efficient energy storage solutions drives the relentless pursuit of next-generation batteries. Conventional lithium-ion batteries, while dominant, face limitations in energy density, safety, and compatibility with certain electrode materials. Ionic liquids (ILs) emerge as a promising class of electrolytes to overcome these hurdles and propel battery technology forward. ILs are salts consisting of organic cations and inorganic/organic anions having melting transitions lower than 100°C. They are an important class of components used as electrolytes for batteries. High ionic conductivity is crucial for battery applications, but traditional ionic liquids (ILs) and monocationic ionic liquid crystals (ILCs) often fall short. This chapter explores a promising alternative: dicationic ionic liquids (DILs) and dicationic liquid crystals (DILCs) based on extended viologens. These materials exhibit remarkably high ionic conductivities ($\sigma_{dc} \approx 10^{-1.5}$ S cm^{-1}), comparable to benchmark electrolytes used in fuel cells and batteries. We report the synthesis of a novel class of DILs and DILCs using a Zincke reaction under mild conditions. These materials feature extended viologen bistriflimides with oligoethyleneoxy groups (DILs) or bistosylates (DILCs). Their excellent ionic conductivity makes them ideal candidates for energy conversion and storage applications. The chapter delves into the detailed synthesis process and the impressive conductivity properties of both DILs and DILCs.

Keywords: extended viologens bistriflimides, oligoethyleneoxy, ionic liquids, zincke salt, ionic liquid crystals, AC conductivity of dicationic ionic liquid

1. Introduction

Ionic liquids (ILs) are low-temperature molten salts, that is, liquids composed of positive (cations) and negative ions (anions). They are typically salts that have the melting points (T_ms) lower than 100°C and even can be lower than room temperature [1]. These materials offer a compelling alternative to conventional solvents like ethylene carbonate and propylene carbonate, showcasing a unique combination of properties. Key features include low vapor pressure, an extensive liquid range, reduced flammability, high polarity, enhanced ionic and thermal conductivity, excellent stability under thermal, chemical, redox, and radiochemical stresses, a wide electrochemical potential window, and the unexpected ability to effectively dissolve a broad spectrum of substances [2]. To date, monocationic ILs that include ammonium, phosphonium, imidazolium, pyridinium, pyrrolidinium, piperidinium, triazolium, among others have been used in many technologies including batteries [3–5]. Recently, another class of ILs known as redox active ionic liquids (RAILs) wherein the redox active moieties including ferrocene, anthraquinone and 2,2,6,6-tetramethylpiperidine-N-oxide are tethered in the cation or anion or both to generate this class of task specific materials. These are touted as potential electrolytes in lithium ion batteries (LIBs) and high-energy supercapacitors because of their high charge densities [6–8]. For the development of solid state batteries, solid polymer electrolytes such as single-ion BAB triblock copolymers can overcome the intrinsic instability of liquid electrolytes that results in safety issues. In addition, they have excellent lithium-ion transport number close to unity, excellent mechanical property and excellent electrochemical stability window spanning 5 V versus Li^+/Li [9].

When employed as electrolyte components, ILs exhibit improved safety and electrochemical performances in lithium-ion batteries. However, the measured ionic conductivities of ILs are not high enough because of their high viscosities that are due to their ionic interactions for commercial applications in batteries [10–15]. Dicationic ionic liquids (DILs) and even multi-cationic ILs have garnered significant global research interest in recent years. This surge in attention stems from the observation that the presence of multiple charges within these compounds can significantly broaden the spectrum of their physical properties. Consequently, enhancements in density, surface tension, and viscosity are often observed, which can be attributed to their increased molecular weights [16–19]. The physical properties of multi-charged ionic liquids (ILs) can be precisely adjusted through the strategic combination of different cations and anions, given their well-defined chemical structures. The large charge densities and electrostatic interactions normally increase the thermal stabilities [20] and electrical capacities [16, 21] of ILs. A critical aspect yet to be fully elucidated is the precise combination of dicationic moieties and anions that optimize DILs for reduced viscosity and, consequently, enhanced conductivity. While current research explores a diverse range of cationic structures, including ammonium, phosphonium, imidazolium, pyridinium, pyrrolidinium, piperidinium, triazolium, and 4,4′-bipyridinium (viologen), the potential for the design and development of novel, tailored materials remain virtually limitless. On the other hand, ionic liquid crystals (ILCs) [22–24] is another class of materials that combine the excellent properties of ILs with those of liquid crystal phases (LCs). Ionic liquid crystals (ILCs) exhibit unique properties that can significantly diverge from those of conventional liquid crystals due to their ionic character. By carefully selecting the components involved in ionic self-assembly, it is possible to design functional ILCs with tailored optical, electrical, magnetic, thermal, structural, or mechanical properties. Depending on their chemical architectures, ILCs often demonstrate high ionic or proton conductivity. This aligns with a key objective in molecular electronics: the development of ion-conductive materials. Notably, the ion conductivity in ILCs can exhibit anisotropy, meaning it varies depending on the direction of measurement. This anisotropic behavior arises from the inherent anisotropic structural organization of ILCs, where ions serve as charge carriers. However, the transport of mobile ions within self-assembled ILCs can be hindered by the combined influence of bulky ions and the viscoelastic nature of the solid-like matrix. These limitations can pose challenges for their effective utilization in electrochemical devices such as capacitors, solar cells, and lithium batteries. Therefore, they can facilitate ion transport by providing pathways through the anisotropic structural organizations in particular through the ionic self-assembly structures. Note here that both properly designed DILs and DILCs having the good ionic conductivities can be used as electrolytes for the next generation 0f batteries including solid state batteries.

In this chapter, we describe the synthesis of four series of different dicationic salts that include DILs and DILCs as potential electrolytes in energy applications, by correlating their ionic conductivities to chemical structural effects.

2. Conductivity measurements

Impedance spectroscopy was employed to investigate the conductivity of the ionic liquids [25]. The molten IL samples were applied to stainless steel electrodes with a spacer in between. The top electrode had a diameter of 10 mm, while the bottom electrode had a diameter of 20 mm. To maintain a consistent 120 μm separation between the electrodes, fused silica fiber spacer with a diameter of 120 μm was utilized. Frequency sweeps were conducted in the range of $f = 0.1$ Hz to 10^6 Hz, and the complex dielectric permittivity and conductivity were measured.

3. Results: Materials synthesis and conductivity measurements

A series of four novel dicationic organic salts has been synthesized and evaluated for their potential as advanced electrolytes in battery systems. The dicationic stilbazolium salts and dicationic asymmetric viologens are synthesized based on general synthetic strategy. The strategy is the quaternization by $S_N{}^2$ reaction aka Menshutkin reaction, followed by metathesis reactions of desired anions [17–21, 26]. Dicationic ionic liquids (DILs) and dicationic ionic liquid crystals (DILCs) are synthesized using Zincke salts as chloride salts, followed by metathesis reactions using lithium of triflimides and silver tosylates, respectively, in a mixture of ethanol and water.

3.1 Dicationic stilbazolium salts (I-1–I-4) and (II-1–II-4): Synthesis and investigation of ionic conductivities

Dicationic stilbazolium salts (**I-1–I-4**) were synthesized via $S_N{}^2$ reaction between *trans*-4-octyloxy-4-stilbazole and a series of a α,ω-methylene ditosylates. Subsequent metathesis reactions with lithium triflimide yielded the corresponding dicationic stilbazolium bistriflimide salts (**II-1–II-4**) (see **Figure 1**) [27]. Detailed synthetic procedures and characterization data, including spectroscopic analyses and elemental analyses, are provided in Ref. [27].

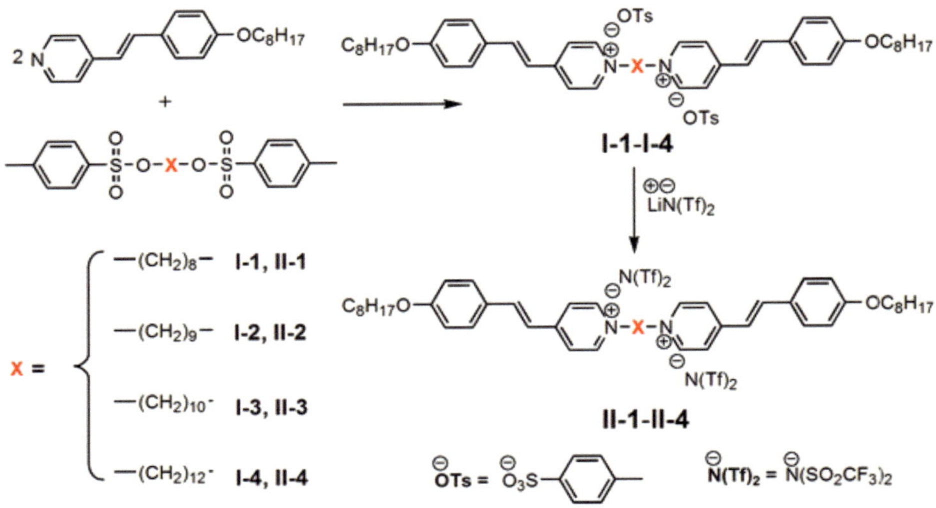

Figure 1.
Synthesis of dicationic stilbazolium dicationic salts (I-1–I-4) and (II-1–II-4) containing bistosylate and bistriflimide counterions.

Our analysis reveals fascinating differences in the dielectric response between tosylate and triflimide salts. As illustrated in **Figure 2(a)**, tosylate salts (**I-*n***) exhibit a stronger and more complex dielectric response compared to their triflimide counterparts (**II-*n***). This can be attributed to the presence of additional aromatic groups in the tosylate structure. Notably, these dielectric relaxations contribute to short-range conductivity within the salts, as evident in **Figure 2(b)**, which highlights the association with local charge displacements. Despite their pronounced dielectric response, tosylate salts surprisingly showed no discernible signs of direct conductivity (σ_{dc}) across the investigated temperature and frequency range. We hypothesize that the bulky aromatic groups within the tosylate structure might impede long-range ionic charge transport. In contrast, triflimide salts exhibit measurable DC conductivity

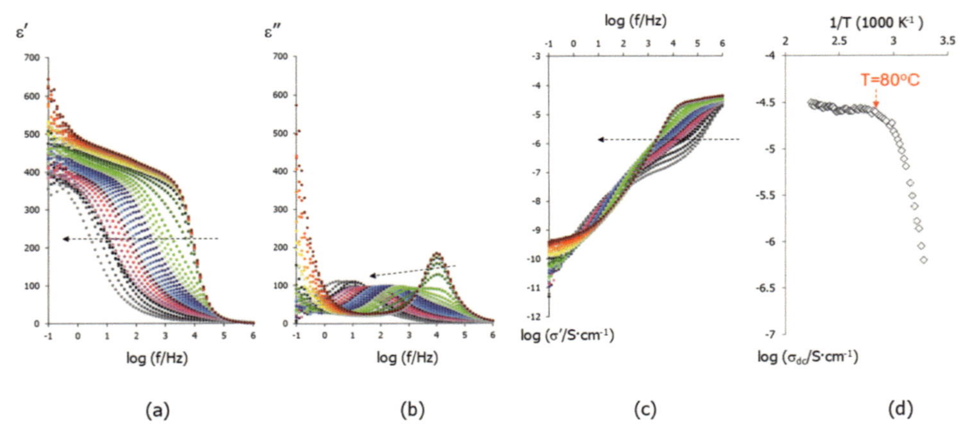

Figure 2.
Complex dielectric permittivity and conductivity plot of the dicationic stilbazolium salts: (a) real dielectric permittivity, ε' (b) imaginary dielectric permittivity, ε'' and (c) real conductivity, σ' of II-3; (d) DC conductivity (σ_{dc}) Arrhenius plots for II-4.

values in the isotropic phase, reaching approximately $10^{-4.5}$ S cm^{-1}. However, a significant decrease in conductivity is observed upon cooling, following a characteristic Vogel-Fulcher-Tammann (VFT) temperature dependence (refer to **Figure 2c**) [27–30].

The strong correlation between conductivity and temperature dependence of ionic diffusion coefficients indicates that effective charge transport in these salts is tightly coupled to local, segmental motions. This observation suggests that maximizing such internal mobility could be crucial for optimizing their electrical performance. The pronounced impact of temperature on conductivity further implies that ionic transport in these materials might not solely rely on direct intermolecular jumps but rather may involve a more convoluted, diffusion-based mechanism facilitated by structural flexibility. The triflimide salt's notable DC conductivity highlighting significant promise for practical battery applications. Their conductivity embodies key characteristics of an ideal battery electrolyte, including high ionic conductivity facilitating swift ion transport between electrodes, thereby enhancing power, and charging speeds. Additionally, these salts possess a wide electrochemical window, enabling stable operation at high voltages and ultimately increasing energy density. Furthermore, their thermal stability across a range of operating temperatures enhances overall safety and reliability in battery systems.

3.2 Dicationic asymmetric viologens, 6BP*n*(s)

The synthetic routes of asymmetric viologens with hexyl terminal group and different alkyl chain lengths are shown in **Figure 3** [31]. The detailed synthetic

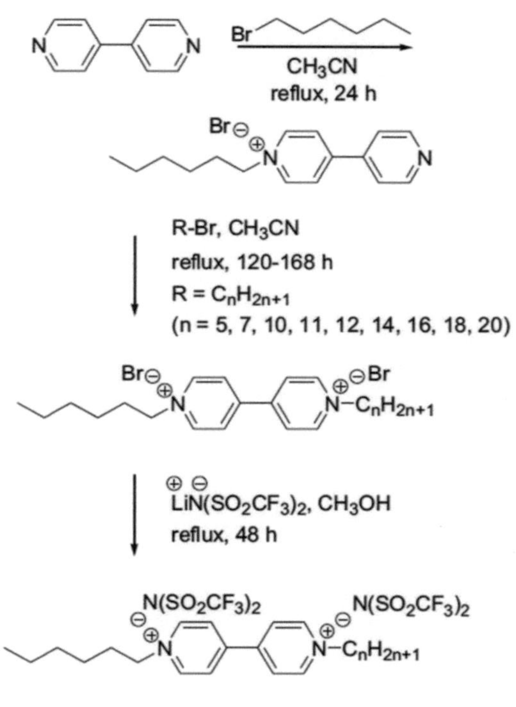

Figure 3.
Synthetic pathways for a series of asymmetric viologen bistriflimide salts derived from a 4,4'-bipyridinium core (BP) featuring hexyl substituents with varying alkyl chain lengths (n = 5, 7, 10, 11, 12, 14, 16, 18, and 20) [30, 31].

procedures and characterizations by spectroscopic techniques and elemental analysis are presented in the reference [31]. The **6BP*n*(s)** exhibit Smectic T phases in a broad range of temperatures, including room temperature, determined by differential scanning calorimetry (DSC) polarized optical microscopy (POM) and variable temperature X-ray diffraction (VT-XRD) studies and belong to the class of ILCs [31].

Figure 4(a) and **(b)** respectively illustrate the dielectric relaxation and high-frequency conductivity of 6BPn(s). The samples exhibit high DC conductivities (σ_{dc}), a characteristic property of Smectic T viologen nanostructures where ionophilic regions are separated by insulating alkyl chains. The observed primary dielectric relaxation in this compound corresponds to the rotation of its molecular core (β-relaxation). High-frequency conductivity, on the other hand, is associated with short-range hopping of triflimide ions. The DC conductivities of these compounds ranging from 10^{-5} to 10^{-3} S cm^{-1}, hold significant promise for applications in organic media. Additionally, a distinct Vogel-Fulcher-Tammann (VFT) behavior was observed in the dielectric processes signifies a strong coupling between the dielectric response, conductivity response, and segmental-type motions of these ionic liquids (ILCs) [30, 31]. This interplay between dielectric relaxation, short-range ion hopping, and DC conductivity in 6BPn(s) opens exciting possibilities for their application in next-generation organic batteries and other electrochemical devices.

3.3 Dicationic extended viologens with bistriflimide counterions (1–3) and their ionic conductivities

The dicationic extended viologens **1–3** are prepared according to the scheme as shown in **Figure 5** using Zincke salts and (4-oligooxyethylene) anilines as chloride salts, followed by metathesis reactions using lithium triflimides in a mixture of ethanol and water [32]. The detailed synthetic procedures and characterizations by spectroscopic techniques and elemental analysis are presented in the reference [32].

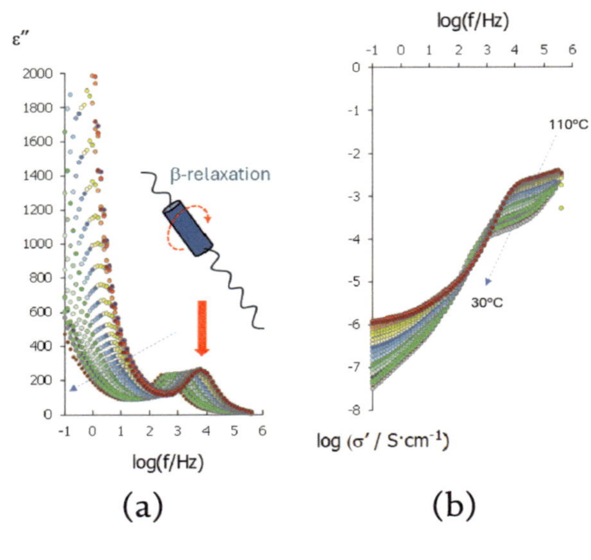

Figure 4.
Complex dielectric results of 6BP11: (a) imaginary permittivity, ε″ ; (b) real conductivity, σ′ .

They are characterized by DSC and POM and found to be ionic liquids. They are the first examples of ionic liquids prepared *via* Zincke reactions [30, 32].

Figure 6 reveals the frequency-dependent dielectric and conductivity response of extended viologens **1–3** as a function of temperature. These salts exhibit remarkably

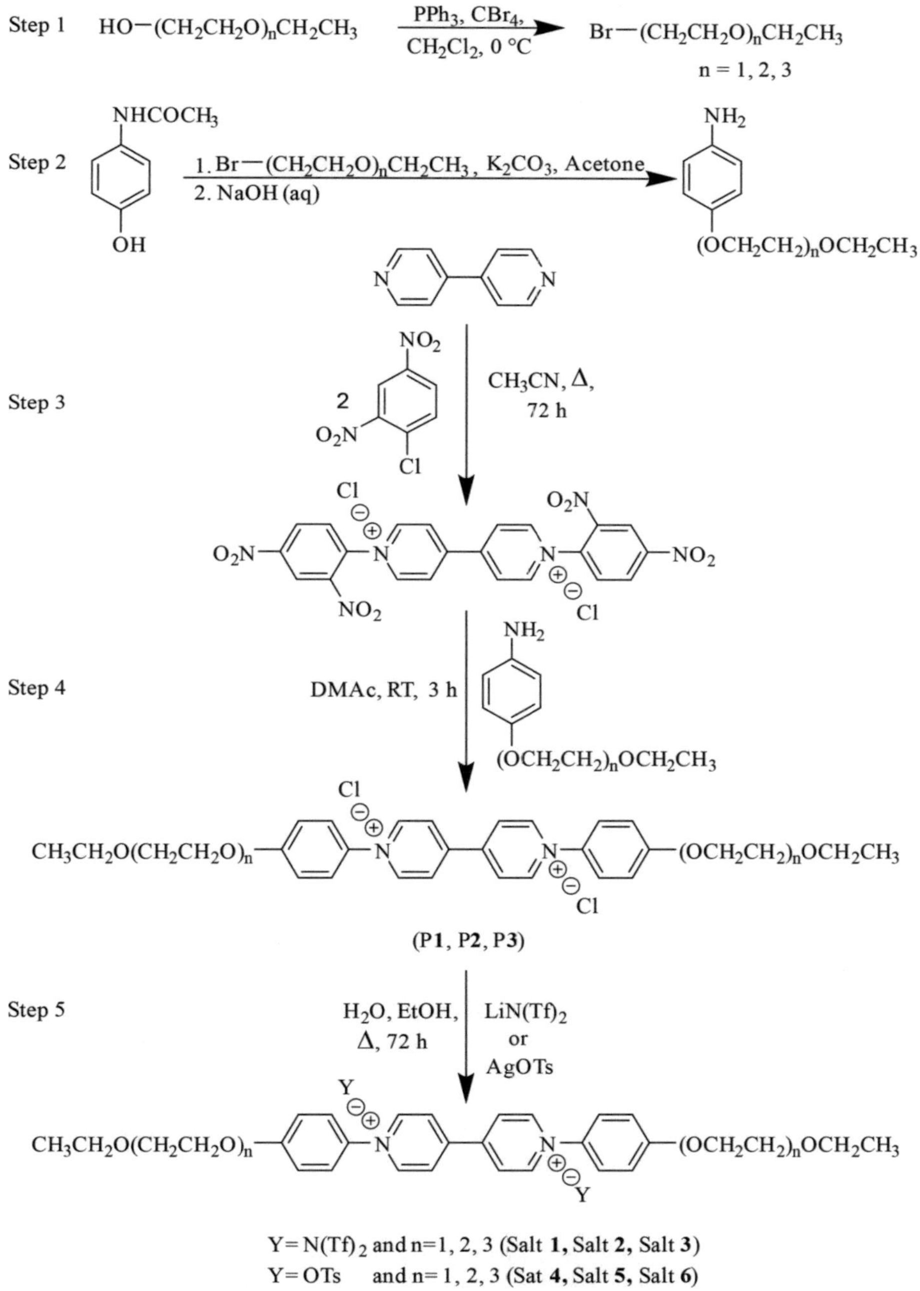

Figure 5.
Synthetic pathways for the preparation of ionic liquids 1–3 and organic salts 4–6.

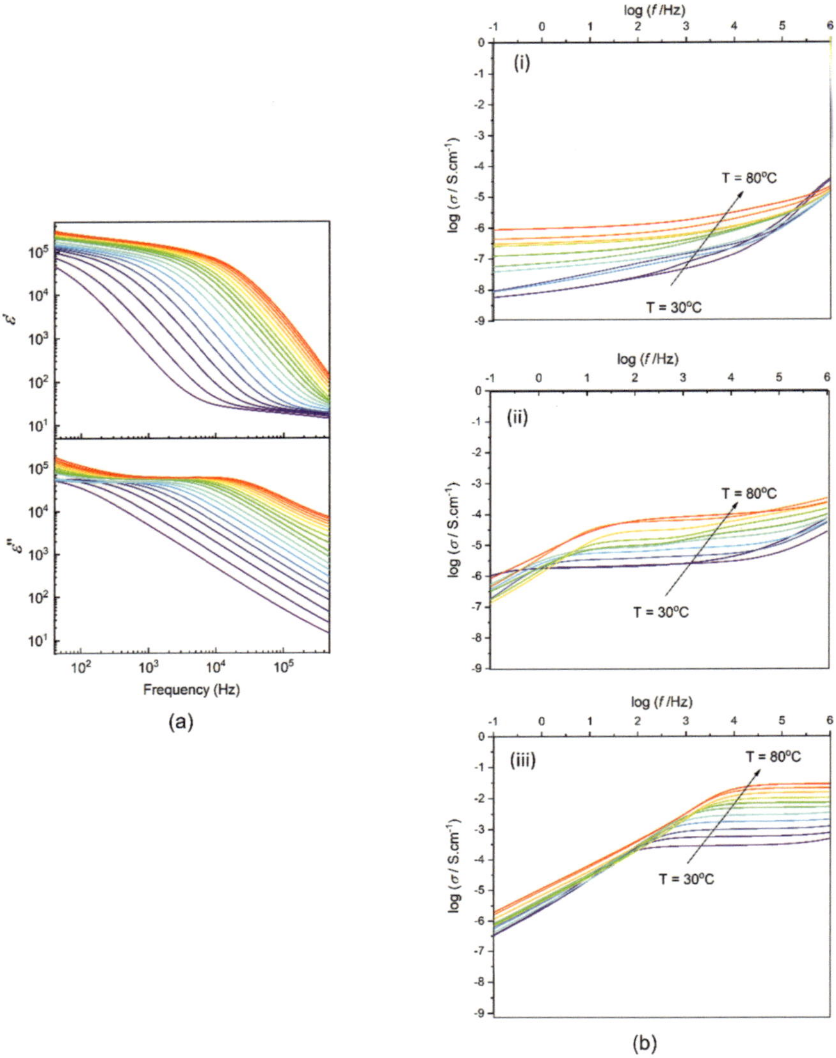

Figure 6.
(a) Complex permittivity of compound 3 as a function of temperature. (b) Conductivity of compounds 1, 2, and 3 plotted against temperature.

high complex permittivity and conductivity values, exceeding those of typical ionic liquids and salts. This exceptional behavior can be attributed to the highly polar nature of the viologen cations [32]. **Figure 6(a)** showcases the frequency-dependent dielectric response of viologen salts **3**. The ε" plot reveals a striking linear increase at lower frequencies, with a slope of −1. This trend signifies a rise in direct current (DC) conductivity, effectively overshadowing any potential dielectric relaxation processes [33]. Salt **3** exhibited the highest ε" values among the three compounds. Further confirmation of DC conductivity comes from the double logarithmic plots of σ' versus f in **Figure 6(b)**. The formation of plateaus in these plots, as described by [25], solidifies the presence of a non-frequency-dependent DC component.

DC conductivity values, σ_{dc}, were estimated by extrapolating the plateau regions of the σ' spectra to zero frequency at various temperatures. Arrhenius plots (not shown) derived from these data were used to determine the activation energy (Ea) of the conduction process. Salt **1** and **2** exhibited activation energies of 95.9 kJ/mol and 84.5 kJ/mol, respectively, suggesting a mechanism involving localized molecular motions, consistent with the occurrence of β-relaxations associated with the rotation of rod-like viologen moieties within the crystal lattice [32, 33]. Conductivity in these salts is notably affected by the length of the terminal chains. Shorter chains, as observed in salts **1** and **2**, seem to impede ionic mobility. This observation suggests that the presence of the bulky four-phenyl core exerts a dominant influence on molecular dynamics, thereby hindering efficient charge transport [34–37]. However, salt **3** breaks the trend. With its longer ethyleneoxy chains ($-(CH_2CH_2O)-$), it exhibits exceptional σ_{dc} conductivity values exceeding 0.1 S cm^{-1}. This puts it on par with established electrolytes used in fuel cells and batteries [38, 39]. This remarkable performance can be attributed to a plasticizing effect, where the longer chains facilitate ionic motion within the material. Furthermore, salt **3** stands out as a rare example of an organic salt with such high conductivity under dry conditions and near room temperature [10, 40–44]. This exceptional behavior can be explained by the formation of a rubbery phase above its low glass transition temperature (−11°C). This molten-like state with large free volume provides ample space for ionic movement, leading to the observed surge in conductivity [45, 46]. In this state, conductivity shows minimal temperature dependence [47–49]. In contrast, the shorter chain salts **1** and **2** remain glassy throughout the studied temperature range, resulting in lower σ_{dc} conductivity values. This highlights the delicate interplay between chain length, molecular dynamics, and ultimately, the efficiency of charge transport within these materials.

3.4 Dicationic extended viologens with bistosylate counterions (4-6) and their ionic conductivities

Dicationic extended viologens with bistosylate counterions **4–6** are synthesized according to the scheme as also shown in **Figure 5** using Zincke salts and (4-oligo-oxyethylene) anilines as chloride salts, followed by metathesis reactions using silver tosylates in a mixture of ethanol and water [50]. The detailed synthetic procedures and characterizations by spectroscopic techniques and elemental analysis are presented in the reference [50]. The salt **4** is a high melting salt that melts to liquid phase. In contrast, the salt **5** and **6** show smectic A phases (SmA) after melting transitions and show SmA to isotropic liquid transitions as determined by differential scanning calorimetry (DSC) polarized optical microscopy (POM) and variable temperature X-ray diffraction (VT-XRD) studies and belong to the class of ILCs [50].

The real component of the complex conductivity, σ, for the three salts (**4–6**), illustrates intriguing trends in **Figure 7**. Notably, conductivity amplifies with the increase of oxyethylene units adjoining the extended viologen moiety. Salt **6**, enriched with four oxyethylene units, showcases the most remarkable direct current conductivity at room temperature (30°C), approximately $\sigma_{dc} \sim 4 \times 10^{-5}$ S cm^{-1}. The elongated oxyethylene chains are likely to enhance molecular flexibility [51], fostering ion movement within the ionic liquid and consequent increase in the conductivity [52]. This aligns with the reduced melting transitions shown in [50]. However, the elongated oxyethylene chains might also escalate the relative concentration of solvation sites [52] and diminish ion-ion interactions [53, 54].

This duality suggests that while extending the end chains offers potential for higher conductivities, it could also induce strong ionic complexation, impeding mobility [55]. Our results in **Figure 8** likely stem from a blend of these factors. The high conductivity of salt **6** makes it a promising candidate for battery applications.

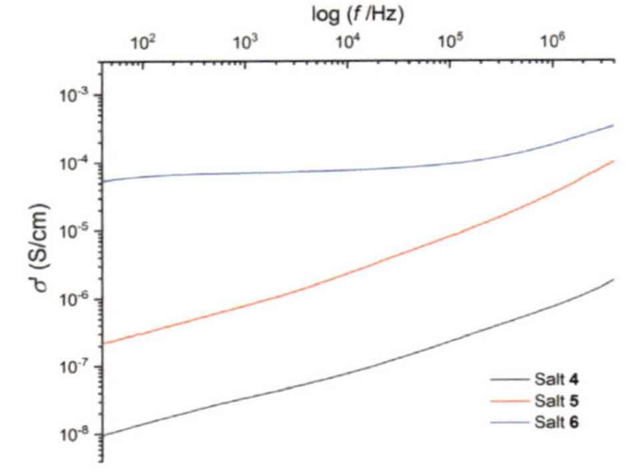

Figure 7.
The real conductivity, σ′ as a function of frequency for the three salts 4–6 at 30°C.

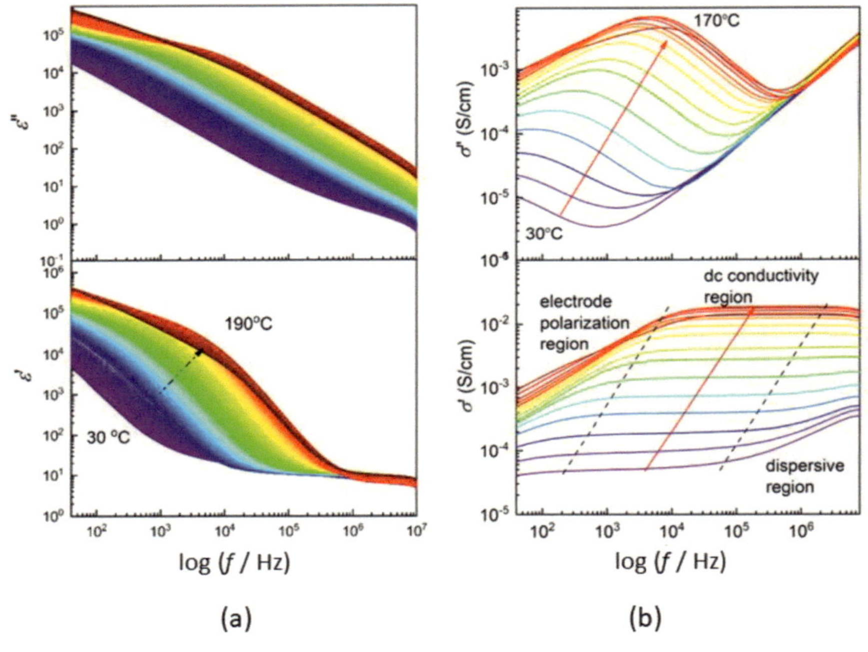

Figure 8.
Temperature dependence of the complex permittivity (ε) of salt 6. (a) Real part (ε′) and imaginary part (ε′′) of the permittivity. (b) Complex conductivity (σ*) of salt 6, showing both real (σ′) and imaginary (σ′) components.*

The response of a material to an applied electric field is captured by its complex permittivity, $\varepsilon^* = \varepsilon' + j\varepsilon''$ [56]. **Figure 8(a)** shows the ε^* of salt **6** obtained by heating from 30 to 190°C. The high ε^* values, as depicted in **Figure 8(a)**, signify a strong dielectric response, associated with the polarizable groups in the molecule.

Figure 8(b) illustrates the temperature-dependent behavior of the complex electrical conductivity of salt **6**. The double logarithmic plots of the real component of the conductivity (σ') versus frequency and temperature unveil three distinct regions. At lower temperatures and high frequencies, conductivity is governed by pronounced dispersion effects, markedly influenced by the frequency of the applied electric field. At intermediate frequencies, ions exhibit sufficient time to hop between electrodes, manifesting as plateaus in σ', indicative of DC conductivity. Finally, at low frequencies, σ' diminishes owing to electrode polarization effects induced by charge accumulation at electrode surfaces [57, 58].

At higher temperatures, the ionic liquid's conductivity exhibits an upward trajectory, with the relaxation peak of ε^* shifting to higher frequencies. This conductivity increase can be ascribed to two probable effects: temperature-induced activation of more ionic sites due to higher molecular mobility and enlarged free volume facilitating ion hopping. Additionally, thermal energy causes the ions increased diffusion, reflected in the shift of the DC conductivity region towards higher frequencies upon heating. The thermal activation energy of conductivity, deduced from Arrhenius plots for salts **4–6**, reveals intriguing insights. While salts **4** and **5** exhibit activation energies of 39 kJ/mol and 49 kJ/mol, respectively, salt **6** displays 57 kJ/mol, despite its highest conductivity. This discrepancy underscores the complex nature of conductivity determinants, where factors such as number of charge carriers and their mobility carry significant influence [59]. Salt **6**, with its longer oxyethylene units, exhibited the highest conductivity at room temperature (as discussed earlier). This is likely due to the increased flexibility of these chains, which enhances ion mobility within the ionic liquid. The observed increase in conductivity for salt **6**, despite its higher activation energy, suggests a more complex transport mechanism. While the exact nature of this mechanism requires further investigation, the increased flexibility of the longer oxyethylene chains may play a crucial role. This increased flexibility could potentially facilitate more efficient ion hopping or reduce the energetic barriers associated with ion transport within the smectic phases. Furthermore, we hypothesize that the formation of more well-defined smectic layers, promoted by the longer oxyethylene chains, may also contribute to enhanced conductivity. The increased order within these smectic phases could provide more efficient pathways for ion migration, leading to improved overall conductivity as evidenced by the data presented in **Figure 9**.

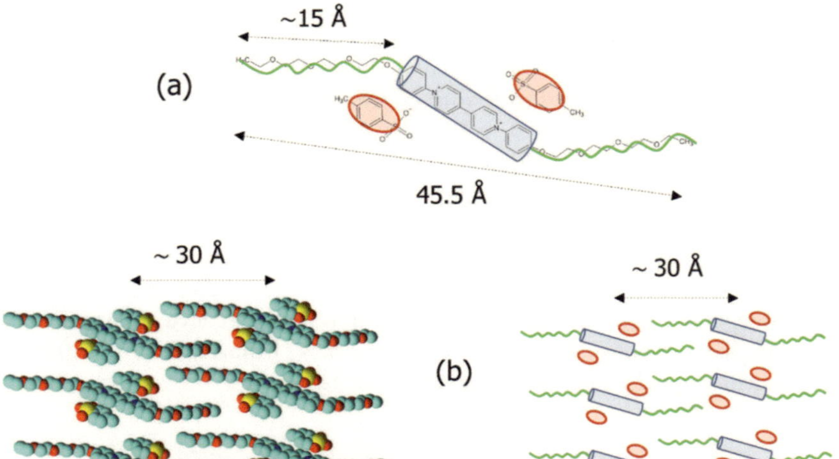

Figure 9.
*(a) Molecular model of salt **6**, including estimated dimensions, generated using ACD Software. (b) Schematic representation of a potential smectic nanostructure for salt **6**, inferred from X-ray diffraction data, highlighting the segregation of aromatic and ethylene oxide regions within the layered structure. Reprinted with permission from Bhowmik et al. 2024. Copyright 2024 Elsevier.*

4. Conclusions

A series of novel viologens were synthesized via Zincke reactions, resulting in the formation of ionic liquids and salts exhibiting strong dielectric properties. Our findings underscore the critical role of a delicate equilibrium between local charge interactions and molecular mobility in governing the phase behavior and conductivity of these ionic liquids. Notably, the formation of amorphous, liquid crystalline, or crystalline phases can be systematically modulated by adjusting the aspect ratio of the rigid core and the flexibility of the terminal functionalities.

While the conductivity values of salts with alkyl terminations remain several orders of magnitude lower than those exhibited by benchmark phosphonium or imidazolium-based ionic liquid electrolytes (10^{-2} S cm^{-1}) [60, 61], our findings demonstrate that the incorporation of oxyethylene groups significantly enhances conductivity, reaching values in the 10^{-1} S·cm^{-1} range, comparable to the requirements for commercial battery and fuel cell applications [38, 39]. The presence of additional polar sites within these terminal groups may be crucial for facilitating long-range ion transport within these materials. Interestingly, the presence of tosylate ions, despite their strong dielectric response, does not appear to promote efficient long-range conductivity. In contrast, triflimide ions seem to be more conducive to ionic transport. These observations suggest that delocalization of charge may be a critical factor in achieving high σ_{dc} values.

Within the series of extended viologens incorporating bistosylate counterions, a notable trend emerges: a gradual increase in electrical conductivity is observed with increasing length of the oxyethylene units within the viologen moiety. This trend culminates in exceptionally high DC conductivities (σ_{dc}) approaching 10^{-2} S cm^{-1}, within the crystalline phase near 100°C. These conductivity values are highly significant, approaching those exhibited by commercially employed electrolytes in cutting-edge fuel cell and battery technologies [38, 39], and surpassing the conductivity levels reported for numerous other liquid crystalline materials [62, 63]. This observation suggests a strong correlation between the length of the oxyethylene spacer and the material's ability to facilitate efficient charge transport within the crystalline lattice.

Our findings demonstrate the potential of these ionic liquids, including other ionic liquids and ILCs, for integration into energy-related devices such as fuel cells, batteries, supercapacitors, and solar cells. The observed trends strongly suggest that the physicochemical properties of these materials, particularly their conductivity, can be significantly influenced by strategic molecular design strategies. For instance, systematically modifying the length of the central rigid core, exploring alternative anionic counterions, and fine-tuning the composition and length of the terminal functionalities offer promising avenues for engineering ionic liquids and ILCs with precisely tailored electrostatic interactions and nanoscale structural organization. This approach has the potential to unlock new classes of materials with optimized properties for high-performance energy applications.

Acknowledgements

PKB sincerely acknowledges the Knowledge Fund that is administered by the Nevada Governor's Office of Economic Development (GOED) and the University of Nevada Las Vegas (UNLV). SLC, SLC, JS, and DK all sincerely acknowledge the financial support from the National Institute of General Medical Sciences (GM103440) of the National Institutes of Health, USA. HH sincerely acknowledges the Faculty Opportunity Award (FOA) that is administered by the Office of Sponsored Programs (OSP) at UNLV. TSV acknowledges Universiti Malaya for the UMREG Grant Scheme [UMREG019-2023] for the financial support. AMF would like to thank the Carnegie Trust for the Universities of Scotland, for the Research Incentive Grant RIG008586, the Royal Society and Specac Ltd., for the Research Grant RGS\R1\201397, and the Royal Society of Chemistry for the award of a mobility grant (M19-0000).

Conflict of interest

The authors declare no conflict of interest.

Author details

Pradip K. Bhowmik[1*], Seonghyeok L. Cox[1], Si L. Chen[1], Jakob Smith[1], David King[1], Haesook Han[1], Thamil Selvi Velayutham[2] and Alfonso Martinez-Felipe[3]

1 Department of Chemistry and Biochemistry, University of Nevada Las Vegas, Las Vegas, NV, USA

2 Low Dimensional Materials Research Center, Department of Physics, Faculty of Science, Universiti Malaya, Kuala Lumpur, Malaysia

3 Chemical Processes and Materials Research Group, School of Engineering, Centre for Energy Transition, University of Aberdeen, King's College, Old Aberdeen, UK

*Address all correspondence to: pradip.bhowmik@unlv.edu

References

[1] Welton T. Ionic liquids: A brief history. Biophysical Reviews. 2018;**10**(3):691-706

[2] Lei Z, Chen B, Koo Y-M, MacFarlane DR. Introduction: ionic liquids. Chemical Reviews. 2017;**117**(10):6633-6635

[3] Eyckens D, Henderson L. A review of solvate ionic liquids: Physical parameters and synthetic applications. Frontiers in Chemistry. 2019;7:263

[4] Brehm M, Pulst M, Kressler J, Sebastiani D. Triazolium-based ionic liquids: A novel class of cellulose solvents. Journal Physical Chemistry B. 2019;**123**(18):3994-4003

[5] Wojcieszak M, Syguda A, Lewandowska A, Marcinkowska A, Siwińska-Ciesielczyk K, Wilkowska M, et al. Synthesis and surface properties of piperidinium-based herbicidal ionic liquids as a potential tool for weed control. Journal of Agricultural and Food Chemistry. 2023;**71**(**11**):4550-4560

[6] Bodin C, Gelinas B, Deng J, Pithaksinsakul K, Zhu Y, Rochefort D, et al. Describing the unsuspected advantage of redox ionic liquids applied to electrochemical energy storage. Current Opinion in Colloid & Interface Science. 2023;**64**:101677

[7] Mourad E, Coustan L, Lannelongue P, Zigah D, Mehdi A, Vioux A, et al. Biredox ionic liquids with solid-like redox density in the liquid state for high-energy supercapacitors. Nature Materials. 2017;**16**:446-454

[8] Gélinas B, Rochefort D. Synthesis and characterization of an electroactive ionic liquid based on the ferrocenylsulfonyl (trifluoromethylsulfonyl)imide anion. Electrochimica Acta. 2015;**162**:36-44

[9] Bouchet R, Maria S, Meziane R, Aboulaich A, Lienafa L, Bonnet J-P, et al. Single-ion BAB triblock copolymers as highly efficient electrolytes for lithium-metal batteries. Nature Materials. 2013;**12**:452-457

[10] Watanabe M, Thomas ML, Zhang SG, Ueno K, Yasuda T, Dokko K. Application of ionic liquids to energy storage and conversion materials and devices. Chemical Reviews. 2017;**117**:7190-7239

[11] Yang QW, Zhang ZQ, Sun XG, Hu YS, Xing HB, Dai S. Ionic liquids and derived materials for lithium and sodium batteries. Chemical Society Reviews. 2018;**47**:2020-2064

[12] Kar M, Plechkova NV, Seddon KR, Pringle JM, MacFarlane DR. Ionic liquids –further progress on the fundamental issues. Australian Journal of Chemistry. 2019;**72**:3-10

[13] Forsyth M, Porcarelli L, Wang X, Goujon N, Mecerreyes D. Innovative electrolytes based on ionic liquids and polymers for next-generation solid-state batteries. Accounts of Chemical Research. 2019;**52**(3):686-694

[14] Yu LP, Chen GZ. Ionic liquid-based electrolytes for supercapacitor and supercapattery. Frontiers in Chemistry. 2019;**7**:272

[15] Tang X, Lv S, Jiang K, Zhou G, Liu X. Recent development of ionic liquid-based electrolytes in lithium-ion batteries. Journal of Power Sources. 2020;**542**:231792

[16] Patil RA, Talebi M, Sidisky LM, Berthod A, Armstrong DW. Gas chromatography selectivity of new phosphonium-based dicationic ionic liquid stationary phases. Journal Separation Science. 2018;**41**(22):4142-4148

[17] Ikeda T. Facile synthesis of tetra-branched tetraimidazolium and tetrapyrrolidinium ionic liquids. ACS Omega. 2021;**6**(30):19623-19628

[18] Jordão N, Cabrita L, Pina F, Branco LC. Novel bipyridinium ionic liquids as liquid electrochromic devices. Chemistry A European Journal. 2014;**20**(14):3982-3988

[19] Maruyama Y, Marukane S, Morinaga T, Honma S, Kamijo T, Shomura R, et al. New design of polyvalent ammonium salts for a high-capacity electric double layer capacitor. Journal of Power Sources. 2019;**412**:18-28

[20] Clarke CJ, Morgan PJ, Hallet JP, Licence P. Linking the thermal and electronic properties of functional dicationic salts with their molecular structures. ACS Sustainable Chemistry and Engineering. 2021;**9**:6224-6234

[21] Matsumoto M, Shimizu S, Sotoike R, Watanabe M, Iwasa Y, Itoh Y, et al. Exceptionally high electric double layer capacitances of oligomeric ionic liquids. Journal of the American Chemical Society. 2017;**139**:16072-16075

[22] Goossens K, Lava K, Bielawski CW, Binnemans K. Ionic liquid crystals: Versatile materials. Chemical Reviews. 2016;**116**(8):4643-4807

[23] Salikolimi K, Sudhakar AA, Ishida Y. Functional ionic liquid crystals. Langmuir. 2020;**36**:11702-11731. DOI: 10.1021/acs.langmuir.0c01935

[24] Kapernaum N, Lange A, Ebert M, Grunwald MA, Haege C, Marino S, et al. Current topics in ionic liquid crystals. ChemPlusChem. 2022;**87**:e202100397. DOI: 10.1002/cplu.202100397

[25] Dyre JC. Some remarks on ac conduction in disordered solids. Journal of Non-Crystalline Solids. 1991;**135**:219-226

[26] Jordão N, Cruz H, Branco A, Pina F, Branco LC. Electrochromic devices based on disubstituted oxo-bipyridinium ionic liquids. ChemPlusChem. 2015;**80**:202-208

[27] Bhowmik PK, Koh JJ, King D, Han H, Heinrich B, Donnio B, et al. Dicationic stilbazolium salts: Structural, thermal, optical, and ionic conduction properties. Journal Molecular Liquids. 2021;**341**:117311

[28] Vogel H. The temperature dependence low of viscosity of fluids. Physikalische Zeitschrift. 1921;**22**:645-646

[29] Fulcher GS. Analysis of recent measurements of the viscosity of glasses-reprint. Journal of the American Ceramic Society. 1992;**75**:1043-1059

[30] Bhowmik PK, Chen SL, Han H, Ishak KA, Velayutham TS, Bendaoud U, et al. High ionic conductivities of ionic materials as potential electrolytes. In: Mutelet F, editor. Industrial Applications of Ionic Liquids. London, UK: Intech; 2022. DOI: 10.5772/intechopen.107949

[31] Bhowmik PK, Noori O, Chen SL, Han H, Fisch MR, Robb CM, et al. Ionic liquid crystals: Synthesis and characterization via NMR, DSC, POM, X-ray diffraction and ionic conductivity of asymmetric viologen bistriflimide salts. Journal Molecular Liquids. 2021;**328**:115370

[32] Bhowmik PK, Chen SL, Han H, Ishak KA, Velayutham TS, Bendaoud U, et al. Dicationic ionic liquids based on bis(4-oligoethyleneoxyphenyl) viologen bistriflimide salts exhibiting high ionic conductivities. Journal of Molecular Liquids. 2022;**365**:120126

[33] Zentel R, Strobl GR, Ringsdorf H. Dielectric-relaxation of liquid-crystalline polyacrylates and polymethacrylates. Macromolecules. 1985;**18**:960-965

[34] Brown AW, Martinez-Felipe A. Ionic conductivity mediated by hydrogen bonding in liquid-crystalline 4-n-alkoxybenzoic acids. Journal of Molecular Structure. 2019;**1197**:487-496

[35] Bojanowski Z, Knapik J, Diaz M, Ortiz A, Ortiz I, Paluch M. Conductivity mechanism in polymerized imidazolium-based ptotic ionic liquid [HSO_3-BVIm] [OTf]: Dielectric relaxation studies. Macromolecules. 2014;**47**:4056-4065

[36] Concellón A, Hernández-Ainsa S, Barberá J, Romero P, Serrano JL, Marcos M. Proton conductive ionic liquid crystalline poly(ethyleneimine) polymers functionalized with oxadiazole. RSC Advances. 2018;**8**:37700-37706

[37] Liang SW, O'Reilly MV, Choi UH, Shiau H-S, Bartels J, Chen Q, et al. High ion content siloxane phosphonium ionomers with very low T-g. Macromolecules. 2014;**47**:4428-4437

[38] Mauritz KA, Moore RB. State of understanding of Nafion. Chemical Reviews. 2004;**104**:4535-4585

[39] Luntz AC, McCloskey BD. Nonaqueous Li-air batteries: A status report. Chemical Reviews. 2014;**114**:11721-11750

[40] Bostwick JE, Zanelotti CJ, Yu D, Pietra NF, Williams TA, Madsen LA, et al. Ionic interactions control the modulus and mechanical properties of molecular ionic composite electrolytes. Journal Materials Chemistry C. 2022;**10**:947-957

[41] Devaki SJ, Sasi R. Ionic liquids/ionic liquid crystals for safe and sustainable energy storage systems. In: Handy S, editor. Progress and Developments in Ionic Liquids. Intech; 2017. pp. 313-336

[42] Liu H, Yu H. Ionic liquids for electrochemical energy storage devices applications. Journal Materials Science and Technology. 2019;**35**:674-686

[43] Papović S, Gadžurić S, Bešter-Rogač M, Vraneš M. Effect of the alkyl chain length on the electrical conductivity of six (imidazolium-based ionic liquids + γ-butyrolactone) binary mixtures. Journal of Chemical Thermodynamics. 2016;**102**:367-377

[44] Xu C, Yang G, Wu D, Yao M, Xing C, Zhang J, et al. Roadmap on ionic liquid electrolytes for energy storage devices. Chemistry An Asian Journal. 2021;**16**:549-562

[45] Angell CA, Imrie CT, Ingram MD. From simple electrolyte solutions through polymer electrolytes to superionic rubbers: Some fundamental considerations. Polymer International. 1988;**47**:9-15

[46] Angell CA. Relaxation in liquids, polymers and plastic crystals-strong fragile patterns and problems. Journal of Non-Crystalline Solids. 1991;**131**:13-31

[47] Vanti L, Alauddin SM, Zaton D, Aripin NFK, Giacinti-Baschetti M, Imrie CT, et al. Ionically conducting and photoresponsive liquid crystalline terpolymers: Towards multifunctional polymer electrolytes. European Polymer Journal. 2018;**109**:124-132

[48] Alauddin SM, Ibrahim AR, Aripin NFK, Velayutham TS, Abou-Zeid OK, Martinez-Felipe A. New side-chain liquid crystalline terpolymers with anhydrous conductivity: Effect of azobenzene substitution on light response and charge transfer. European Polymer Journal. 2021;**146**:110246

[49] Alauddin SM, Aripin NFK, Velayutham TS, Martinez-Felipe A. Liquid crystalline polymers containing sulfonic and light-responsive groups: From molecular design to conductivity. Molecules. 2020;**25**:2579

[50] Bhowmik PK, Chen SL, King D, Han H, Bolyard Z, Schmiedl C, et al. Dicationic ionic liquid crystals: Synthesis, characterization, optical properties, and ionic conductivity of bis(4-oligoethyleneoxyphenyl) viologen bistosylate salts. Journal of Molecular Liquids. 2024;**398**:124140

[51] Choi UH, Price TL Jr, Schoonover DV, Xie R, Gibson HW, Colby RH. Role of chain polarity onion and polymer dynamics: Molecular volume-based analysis of the dielectric constant for polymerized norbornene-based ionic liquids. Macromolecules. 2020;**53**(23):10561-10573. DOI: 10.1021/acs.macromol.0c02100

[52] Gao T, Itliong J, Kumar SP, Hjorth Z, Nakamura I. Polarization of ionic liquid and polymer and its implications for polymerized ionic liquids: An overview towards a new theory and simulation. Journal of Polymer Science. 2021;**59**(21):2434-2457. DOI: 10.1002/pol.20210330

[53] Bruce PG, Vincent CA. Polymer electrolytes. Journal of the Chemical Society, Faraday Transactions. 1993;**89**(17):3178-3203. DOI: 10.1039/FT9938903187

[54] Jiang HJ. Structural design of ionic liquids for process optimization [Ph.D. dissertation]. Australia: The University of Sydney; 2018

[55] Meyer WH. Polymer electrolytes for lithium-ion batteries. Advanced Materials. 1998;**10**(6):439-448. DOI: 10.1002/(SICI)1521-4095(199804)10:6<C439::AID-ADMA439>E3.0.CO;2-I

[56] Barsoukov E, Macdonald JR. Impedance Spectroscopy. 2nd ed. New York: Wiley; 2005

[57] Wubbenhorst M, van Turnhout J. Analysis of complex dielectric spectra. I. One-dimensional derivative techniques and three-dimensional modeling. Journal of Non-Crystalline Solids. 2002;**305**:40-49. DOI: 10.1016/S0022-3093(02)01086-4

[58] van Turnhout J, Wübbenhorst M. Analysis of complex dielectric spectra. II. Evaluation of the activation energy landscape by differential sampling. Journal of Non-Crystalline Solids. 2002;**305**:50-58. DOI: 10.1016/S0022-3093(02)01120-1

[59] Bloom H, Heymann E. The electric conductivity and the activation energy of ionic migration of molten salts and their mixtures. Proceedings of the Royal Society London A Mathematical and Physical Sciences. 1947;**188**(1014):392-414. DOI: 10.1098/rspa.1947.0016

[60] Jourdain A, Serghei A, Drockenmuller E. Enhanced ionic conductivity of a 1,2,3-triazolium-based poly(siloxane ionic liquid) homopolymer. ACS Macro Letters. 2016;**5**:1283-1286

[61] Maximean DM, Cîrcu V, Paul Ganea CP. Dielectric properties of a bisimidazolium salt with dodecyl sulfate

anion doped with carbon nanotubes. Beilstein Journal of Nanotechnology. 2018;**9**:164-174

[62] Kato T, Yoshio M, Ichikawa T, Soberats B, Ohno H, Funahashi M. Transport of ions and electrons in nanostructured liquid crystals. Nature Review Materials. 2017;**2**(4):17001. DOI: 10.1038/natrevmats.2017.1

[63] Alauddin SM, Aripin NFK, Velayutham TS, Martinez-Felipe A. Liquid crystalline copolymers containing sulfonic and light-responsive groups: From molecular design to conductivity. Molecules. 2020;**25**(11):2579. DOI: 10.3390/molecules25112579

Chapter 5

Flexible Polymerized Ionic Liquids Gel Polymer Electrolytes for Supercapacitor Application

Debalina Deb

Abstract

Supercapacitors are garnering increasing attention due to their rapid charge/discharge cycles, long cycle life, and high-power density. However, their performance is significantly limited by the choice of electrolyte. Ionic liquids (ILs) offer a promising alternative with their wide electrochemical window, high thermal stability, and non-flammability. However, their high viscosity and limited ionic conductivity pose challenges. Flexible polymerized ionic liquids gel polymer electrolytes (PIL-GPEs) emerge as a solution, combining the advantages of both ILs and solid polymer electrolytes (SPEs). PIL-GPEs inherit the high ionic conductivity and wide potential window of ILs, while the polymer matrix provides mechanical stability and prevents leakage. Additionally, their flexibility makes them suitable for next-generation wearable and flexible electronics. This review delves into the recent advancements in PIL-GPEs for supercapacitor applications. It explores the synthesis methods, material properties, and electrochemical performance of various PIL-GPEs. Additionally, challenges and future directions for research are discussed.

Keywords: supercapacitors (SCs), ionic liquids (ILs), polymerized ionic liquids (PILs), gel polymer electrolytes (GPEs), flexible electronics, electrochemical performance, energy storage devices

1. Introduction

The escalating energy crisis and dwindling fossil fuel reserves drive researchers to seek efficient energy storage solutions. Supercapacitors (SCs), also known as ultracapacitors, have emerged as promising candidates to address this pressing need, particularly in the context of rapidly evolving portable electronics and hybrid vehicle technologies [1, 2]. The desirable characteristics of an ideal SC include high output performance (defined by power density, rate capability, and cycle life), exceptional cyclability, and environmental sustainability [3–5]. Based on their energy storage mechanism, SCs fall into two categories: electric double-layer capacitors (EDLCs) and pseudocapacitors [3, 6]. EDLCs typically utilize porous carbon materials like graphene [7] or heteroatom-doped graphene (e.g., N-doped, B-doped, Fe-doped, S-doped) [8, 9], while pseudocapacitors primarily rely on binary, ternary, or even quaternary metal oxides and hydroxides [10–12]. From a fabrication perspective, SCs can be classified as either symmetric (with two identical electrodes) [13] or asymmetric (with two dissimilar electrodes) [14]. Asymmetric supercapacitors (ASCs) employing a battery-type positive electrode and a capacitor-type negative electrode have gained significant traction due to their expanded potential window, leading to enhanced energy density (E) and power density (P).

In recent decades, room-temperature ionic liquids (RTILs) have attracted considerable attention due to their remarkable and diverse properties. These properties include exceptional thermal and chemical stability, the ability to tune their structure across a broad operating temperature range, a wide electrochemical stability window, high ionic conductivity within the range of 10^{-3}–10^{-2} S/cm at room temperature, and non-flammability [15], rendering them promising candidates for Electrochemical Energy Storage (EES) devices such as Lithium-Ion Batteries (LIBs) [16], electric double-layer Supercapacitors (SCs) [17, 18], proton exchange membrane fuel cells [17, 18], and solar cells [19, 20].

Utilizing ionic liquids (ILs) as an alternative to organic electrolytes offers advantages like enhanced ion mobility and eliminating hazards associated with organic electrolytes. Furthermore, ILs often exhibit comparable zero or negligible vapor pressure at normal temperatures owing to their high thermal stability. Consequently, ILs have become extensively employed as solvents or electrolytes in recent times for energy storage applications [16–24].

Typically categorized as organic salts or molten salts, ILs have lower melting points (<100°C) and a wide degree of structural variability. They consist of organic cations, including pyridinium (PY) [25], imidazolium (Im) [15], pyrrolidinium (PYR) [16], ammonium [26, 27], phosphonium [27, 28], and sulfonium [29], derivatives, paired with inorganic or organic anions such as Cl^-, BF_4^-, PF_6^- [30–32], triflate ($CF_3SO_3^-$) [33], bis(fluorosulfonyl)imide (FSI) ($(CFSO_2)_2N^-$) [28, 34], and bis(trifluoromethanesulfonyl imide) (TFSI) ($(CF_3SO_2)_2N^-$) [15, 16]. The various combinations of cations and anions unveil numerous possibilities for tailoring ILs with compounds suitable for specific applications.

Despite their intriguing properties, ILs face several challenges that impede their widespread practical use in supercapacitors. One of the foremost obstacles is the high cost associated with the production of many ILs, rendering them economically less viable for large-scale applications. Moreover, ILs often exhibit higher viscosity compared to traditional electrolytes, leading to increased internal resistance and diminished supercapacitor performance [35, 36]. While some ILs display high ionic conductivity, they may not reach levels sufficient for optimal supercapacitor functionality, particularly when compared to other commercially available electrolytes. Additionally, the electrochemical stability window of certain ILs may be limited, constraining the operational voltage range of supercapacitors [13]. The compatibility of ILs with various electrode materials is crucial for performance, and certain ILs may not be suitable for use with specific materials. Environmental concerns also come into play, as some ILs contain toxic components, raising questions about their overall environmental impact [37]. Furthermore, the commercial availability of certain ILs is restricted, limiting their practical adoption. Researchers are actively working to address these challenges through the development of new IL formulations and exploring ways to enhance their compatibility with supercapacitor systems. While ILs have not yet achieved widespread usage in commercial supercapacitors, ongoing research efforts hold promise for overcoming these limitations and establishing ILs as practical components in future supercapacitor technologies. One such intriguing development involves the incorporation of ionic liquids (ILs) directly into the polymer backbone, creating materials with a captivating dual nature: acting as both electrolyte and separator simultaneously [24, 38–40].

Polymerized Ionic Liquids Gel Polymer Electrolytes (PIL-GPEs) are increasingly recognized as superior options for supercapacitors due to their exceptional properties. The polymerization of ionic liquids into gel polymer electrolytes results in materials with high ionic conductivity, ensuring efficient ion movement within the electrolyte [41]. The gel structure imparts mechanical strength, enhancing the overall stability of the supercapacitor. PIL-GPEs exhibit flexibility and processability, allowing for easy integration into various supercapacitor designs [42]. They offer wider electrochemical stability windows, enabling operation at higher voltages and contributing to increased energy density [43]. The gel structure also reduces internal resistance, leading to improved charge/discharge rates and high-power performance [44, 45]. Moreover, PIL-GPEs enhance safety by minimizing flammability and volatility. Their compatibility with various electrode materials provides versatility in supercapacitor design. Environmentally friendly formulations further align with sustainability goals. Altogether, these attributes position PIL-GPEs as promising materials, addressing key challenges and contributing to the advancement of high-performance, safe, and environmentally friendly supercapacitors. Ongoing research aims to optimize these materials for practical and efficient supercapacitor applications.

While PIL-GPEs offer several advantages for supercapacitor applications, certain limitations need consideration [39, 46–48]. One significant challenge is associated with the cost of ionic liquids, which can be relatively high. This factor can impact the overall cost-effectiveness of PIL-GPEs in large-scale supercapacitor manufacturing. Additionally, the polymerization process and synthesis of PIL-GPEs can be complex, requiring careful optimization for large-scale production. The high viscosity of ionic liquids, even when polymerized, may contribute to increased internal resistance within the supercapacitor, potentially limiting its overall performance. While efforts are made to address environmental concerns by designing eco-friendly PIL-GPEs, the toxicity of certain ionic liquids may still pose challenges in terms of safety and sustainability. Furthermore, the compatibility of PIL-GPEs with specific electrode materials may need further exploration [49, 50], as the effectiveness of these electrolytes can vary with different electrode configurations. Overall, addressing these limitations is crucial for realizing the full potential of PIL-GPEs in practical supercapacitor applications and ensuring their competitiveness with other electrolyte materials. Ongoing research endeavors aim to overcome these challenges and optimize the performance of PIL-GPEs for broader implementation in energy storage devices.

This review is dedicated to highlighting recent advancements in the realm of PIL-based electrolytes, specifically focusing on their potential applications in Supercapacitors (SCs). The discussion encompasses various strategies, opportunities, and challenges associated with the rational design of PIL structures, aiming to enhance their suitability for the fabrication of environmentally friendly Electrical Energy Storage (EES) devices. Emphasis is placed on the exploration of physicochemical and electrochemical properties in selectively modified electrolytes incorporating PILs, contributing to their efficacy in energy storage systems.

2. Synthesis methods for PIL-GPEs

2.1 Types and methods of synthesis of polymer-ionic liquid (PIL) electrolytes

Three primary synthetic routes have been established for the PIL preparation (**Figure 1**).

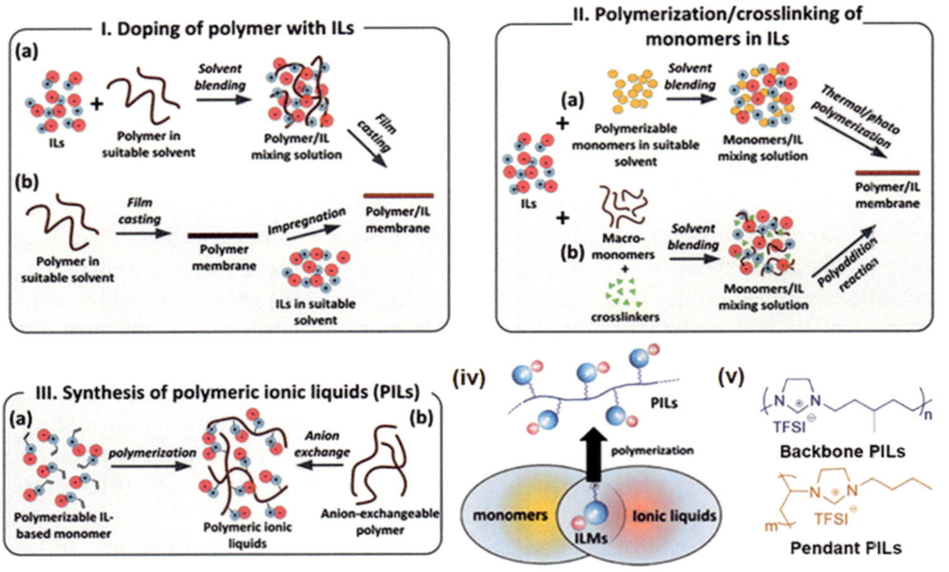

Figure 1.
(i)–(iii) Schematic representation showing the preparation of IL-based polymer electrolytes (reprinted with permission from Ref. [51]). (iv) and (v) Illustration of the structure of poly(ionic liquid)s (PILs) and molecular structures of backbone PIL and pendant PIL (reprinted with permission from Ref. [52]).

i. Doping of polymers with ionic liquids (ILs):

This approach involves incorporating discrete IL molecules into pre-existing polymer matrices. The design focuses on optimizing the compatibility and interaction between the polymer and IL to achieve the desired ionic conductivity and mechanical properties.

ii. In-situ polymerization or cross-linking of monomers in ILs:

This strategy employs ILs as reaction media and templates for the polymerization or cross-linking of monomers. The resulting polymer network encapsulates the ionic moieties within its structure, leading to enhanced ionic conductivity and potential control over morphology.

iii. Synthesis of polymeric ionic liquids (PILs):

Unlike the first two methods that utilize separate polymer and IL components, this approach focuses on the direct synthesis of PILs. These functional polymers possess ionic groups integrated into their main chain or side chains, mimicking some of the key characteristics of ILs, such as high ionic conductivity and thermal stability.

Polymeric ionic liquids (PILs) can be categorized based on the location of their ionic moieties within the polymer chain. As depicted in **Figure 1(iv)**, backbone PILs integrate ionic groups directly into the main chain, while pendant PILs possess them as side groups. Regardless of the PIL type, ionic conductivity arises from the free movement of these charged species. PILs hold promise as polymer hosts for ionogel electrolytes in supercapacitors. Compared to traditional polymers, PIL-based ionogels offer several advantages as solid electrolytes for supercapacitors. These include superior miscibility between PILs and ionic liquids (ILs), facilitating ion transport, and strong interactions between ions and the PIL host, potentially leading to enhanced electrochemical performance.

Polymeric ionic liquids (PILs) come in a variety of structures depending on their building blocks and final design. One way to categorize them is based on the charge of their ionic groups. Polycationic PILs (A) have positively charged groups lining their backbone, while polyanionic PILs (B) have negatively charged groups. Copolymers (C) combine both positive and negative charges within the same polymer chain. Network PILs (D) differ in form, creating a web-like structure through cross-linking for enhanced stability. Finally, poly(zwitterionic) PILs (E) are unique, containing both positive and negative charges within a single repeating unit, yet remaining overall neutral. This diversity in structure allows scientists to tailor PIL properties for specific applications (**Figure 2**).

Now to moving forward to synthesis with the emerging and promising electrolyte—i.e., PIL-GPEs, there are various methods that have been developed, each offering its advantages and limitations. Some common methods include:

2.2 Ex-situ polymerization method

Ex-situ methods offer efficient strategies for preparing polymeric electrolytes (PEs) for practical devices. These methods involve fabricating the PE outside the cell and then integrating it with the electrode and separator assembly; some of the common techniques are:

1. Solution casting: This common method involves dissolving the polymer and lithium salt in an organic solvent to form a blended solution. This solution is then cast onto a flat substrate and dried, resulting in a solid-phase PE film [53, 54].

2. Phase inversion/separation: This technique facilitates the fabrication of porous membranes. The polymer is dissolved in an organic solvent and cast onto a substrate. The substrate is then immersed in a nonsolvent liquid (e.g., water or ethanol), triggering phase separation. The organic solvent diffuses into the nonsolvent, while the nonsolvent penetrates the polymer. Subsequent drying removes the nonsolvent and leaves behind a porous polymer network. These pores can readily absorb liquid electrolytes (LEs) to form gel polymer electrolytes (GPEs) [55, 56].

3. Electrospinning: Electrospinning emerges as a powerful tool for crafting gel polymer electrolytes (GPEs) in batteries and supercapacitors. Its significance lies in weaving intricate fibrous membranes with high porosity and surface area, boosting Li^+ transport and electrolyte uptake [57–59]. By fine-tuning polymer choice and

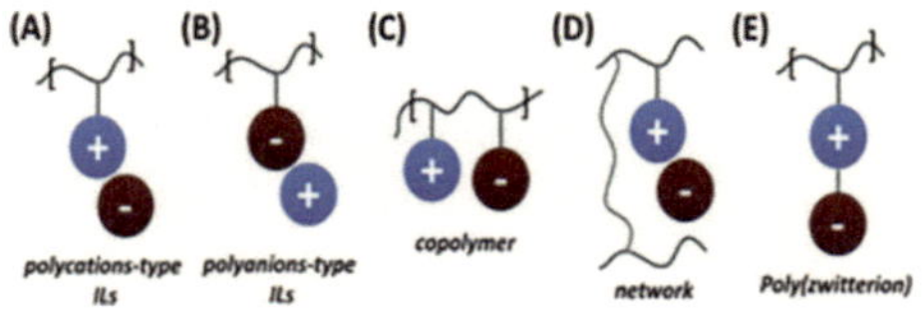

Figure 2.
Different types of PIL (reprinted with permission from Ref. [51]).

electrospinning parameters like voltage and flow rate, we can sculpt the pore size, thickness, and even fiber alignment of these membranes, granting us exquisite control over the GPE's properties. This tailored nanofiber architecture translates to enhanced ionic conductivity, paving the way for high-performance batteries. Scalability also shines as strength, making electrospinning a potential champion for mass production. However, choosing the right polymer blend and optimizing the electrospinning parameters remain challenges to overcome. Additionally, ensuring compatibility between the polymer and the chosen liquid electrolyte is crucial for optimal performance and longevity. Despite these hurdles, electrospinning's unique capabilities hold immense promise for revolutionizing GPE design and unlocking the full potential of next-generation energy storage devices.

While ex-situ methods offer ease of fabrication and control, they face critical limitations in battery applications:

- Interfacial contact: PEs prepared ex-situ typically exhibit imperfect contact with the anode and cathode electrodes, leading to increased interfacial resistance. This resistance hinders Li^+ transport and reduces cycle performance [50].

- Limited electrode utilization: Ex-situ PEs often contact the electrodes primarily on their upper surfaces, neglecting the deeper regions (**Figure 3a**). Consequently, Li^+ transport across the entire electrode is hampered, causing underutilization of active materials and capacity loss.

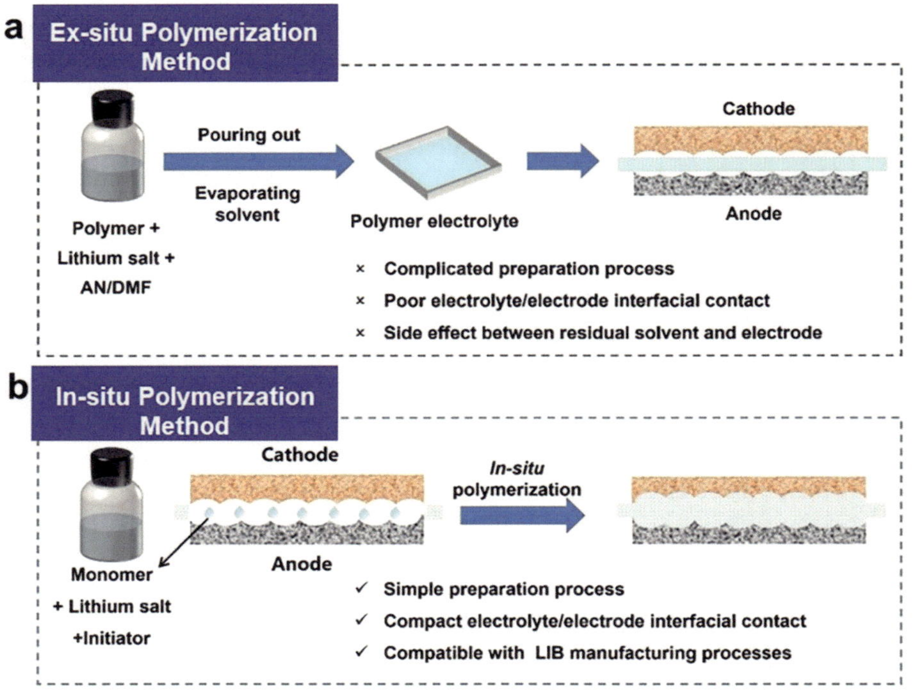

Figure 3.
Schematic representation of (a) ex-situ polymerization method, and (b) in-situ polymerization methods followed by the advantages and disadvantages. (reprinted with permission from Ref. [60]).

- Dendrite growth: Attempting to mitigate interfacial resistance by inserting thin electrolyte films introduces another risk. These films can be susceptible to damage and may not effectively suppress lithium dendrite growth, jeopardizing battery safety.

The ex-situ methods offer efficient PE fabrication; their limitations regarding interfacial contact, electrode utilization, and dendrite growth pose significant challenges for high-performance batteries and supercapacitors. Exploring alternative approaches, such as in-situ polymerization, could mitigate these issues and improve overall battery performance and safety.

2.3 In-situ polymerization

The limitation suffered from interfacial issues and limited electrolyte uptake within the electrode pores by employing ex-situ techniques can be overcome by employing in-situ polymerization, where a liquid precursor solution is directly injected into the electrode and subsequently polymerized using thermal or photochemical methods resulting in a covalently linked polymer network [61, 62].

The liquid precursor readily infiltrates the internal pores of the electrode, resulting in PEs both within the pores and on the active material's surface, as shown in **Figure 3b**. This intimate contact between the electrode and the in-situ polymerized PEs fosters facile Li^+ transport and significantly reduces the interfacial resistance [63].

2.3.1 Composition of in-situ polymerization

The precursor for in-situ polymerization typically consists of four main components:

1. *Monomers with functional groups*: These reactive groups, often acrylates or methacrylates, enable cross-linking during polymerization. Common examples include trimethylolpropane ethoxylate triacrylate (ETPTA), poly(ethylene glycol) diacrylate (PEGDA), trimethylolpropane propoxylate triacrylate (TPPTA), and di(trimethylolpropane) tetraacrylate (DTPTA) [64–66].

2. *Ionic salt*: This provides mobile Li^+ or Na^+ ions for ionic conductivity within the PE.

3. *Solvent*: This dissolves the other components and facilitates effective penetration of the precursor into the electrode.

4. *Initiator*: Thermal or photo-initiators trigger the polymerization reaction upon activation. Radical initiators, either carbo-radicals or oxy-radicals, attack the double bonds of the monomer or the reactive functionalities in the presence of light or heat, facilitating cross-linking and network formation of the polymer matrix (**Figure 4**) [68, 69].

2.3.2 Mechanism of in-situ polymerization

In-situ polymerization within the cells commonly employs two initiation methods: thermal and ultraviolet (UV) light photo-illumination. In thermal polymerization, azobisisobutyronitrile (AIBN) acts as a radical initiator. Upon heating, it

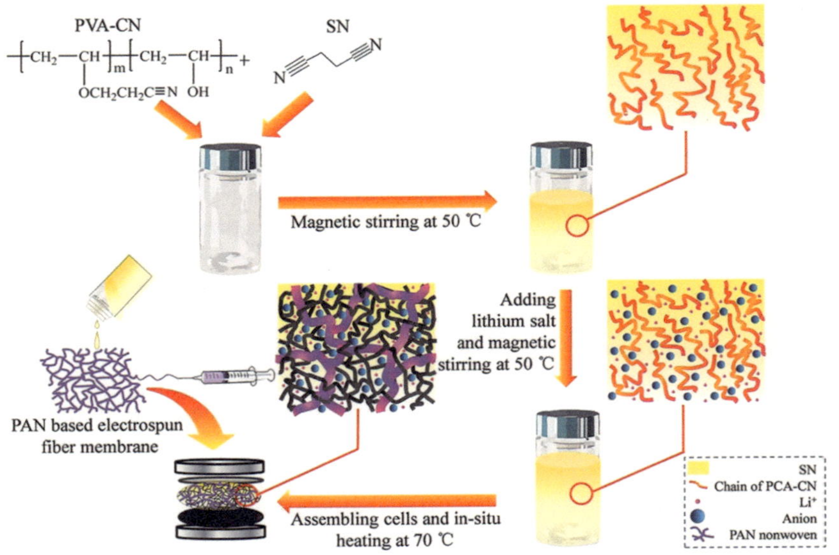

Figure 4.
Schematic illustration for in-situ polymerization (reprinted with permission from Ref. [67]).

decomposes into highly reactive radicals that abstract hydrogen atoms from suitable monomers, creating new radicals. These new radicals then initiate a chain reaction where they repeatedly add monomer units to their growing chains, ultimately forming the desired polymer. The rate of polymerization depends on factors like temperature, AIBN concentration, and monomer reactivity, with the process ending when growing chains encounter other radicals in a termination reaction. Whereas photo-initiators in polymerization act as light-activated catalysts. Upon absorbing specific light wavelengths, these molecules undergo chemical transformations, generating reactive free radicals (for radical polymerization) or Lewis acids (for cationic polymerization). These reactive species then attack and activate monomer molecules, initiating a chain reaction where monomers add themselves repeatedly, ultimately building the desired polymer chain. The choice of photo-initiator and light source depends on the type of polymerization needed, desired speed, and other factors like material compatibility.

Each method offers distinct advantages and drawbacks; for example, with thermal initiation, the thermal energy readily diffuses and ensures sufficient polymerization throughout the electrolyte even after complete assembly and sealing. However, the main issue with thermal initiation is the extra time requirement; it takes around 20–30 minutes for adequate polymerization, potentially leading to component degradation, and increased manufacturing lead times. In contrast, UV light photo-illumination comes with the advantage of complete polymerization within 10 seconds, which significantly reduces the processing time and improves production efficiency. But this comes with complex manufacturing; for example, UV illumination within the sealed cells adds complexity to fabrication and handling procedures. Hence, the choice between thermal and UV initiation depends on the specific priorities and limitations of the battery design and manufacturing process. While thermal initiation offers improved polymer distribution, its potential negative effects on cell components and production time must be carefully considered. Conversely, UV

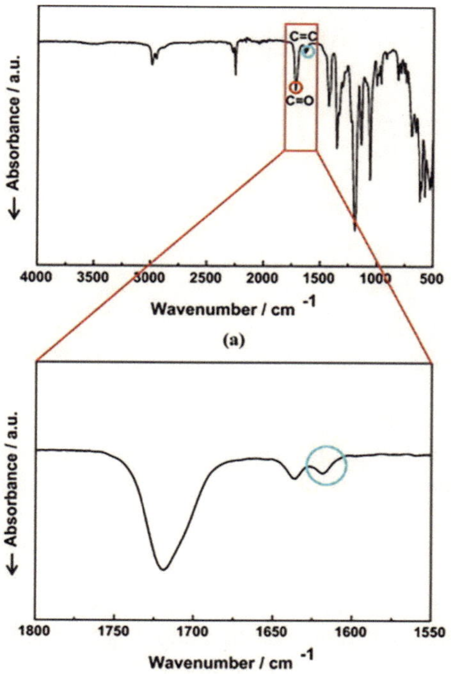

Figure 5.
FTIR spectra of acrylic C=C double bonds (a) before polymerization, (b) after polymerization (reprinted with permission from Ref. [69]).

photo-illumination provides rapid curing but necessitates additional design and handling constraints due to its limited penetration depth.

Fourier-transform infrared (FT-IR) or Raman spectroscopy can be employed to confirm the successful in-situ polymerization. Notably, the disappearance of the characteristic C=C peak (**Figure 5**) in the precursor spectrum within the gel polymer electrolyte (GPE) confirms the completion of the reaction [70, 71].

3. Material properties and electrochemical performance

The current challenge for supercapacitors lies in improving their specific energy while maintaining high specific power. Currently, organic electrolytes dominate SCs, but their volatility, limited conductivity, and toxicity pose concerns. Ionic liquid-based gel electrolytes emerge as a promising alternative offering several advantages: high ionic conductivity, non-flammability, non-volatility, and a wide operational voltage window. This versatility allows ionic liquids to function directly as liquid electrolytes, as salts in organic solvents, or integrated into solid polymer electrolytes. Common anions in ionic liquids include $TFSI^-$, BF_4^-, and PF_6^-, while cations are predominantly large organic species like imidazoles, pyrroles, and short-chain fatty quaternary amine salts. Notably, **Figure 6** demonstrates the potential for wide operational voltage windows of specific ionic liquids at different electrodes, highlighting their compatibility with high-voltage applications. In essence, ionic liquid-based gel electrolytes present a compelling avenue for enhancing the specific energy and expanding the operational potential of SCs, thereby addressing the current limitations of the technology.

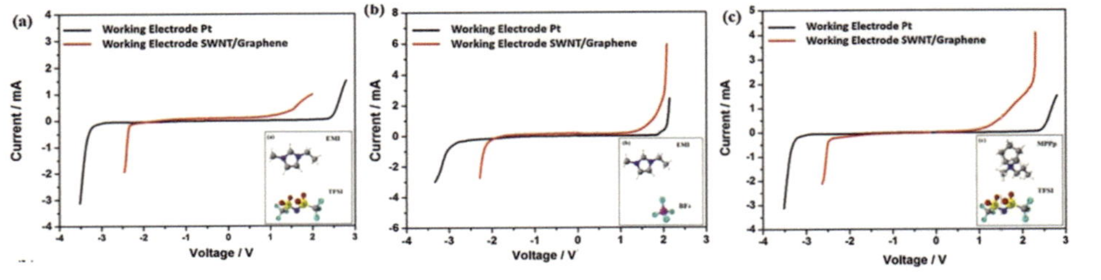

Figure 6.
Electrochemical stability window measurement (using the three-electrode system) with working electrode Pt (colored black) and SWNT/graphene (colored red) in (a) EMI-TFSI, (b) EMI-BF_4, and (c) MPPp-TFSI (the inset of each figure shows the chemical structure of respective ionic liquid) (reprinted with permission from Ref. [72]).

ILs can be used as supporting electrolytes as they can provide mainly cations and anions in organic solution just like conventional salts for charge transport; because of these main reasons, the researchers have used them to widen the working voltage window and thus increase the energy density of the storage systems. In the quest for eco-friendly energy solutions, PVA-based IL gel electrolytes emerge as a promising candidate. Not only do they inherit the superior properties of ionic liquids, like broad voltage windows and enhanced energy density, but they also boast impressive green credentials. PVA itself is non-toxic, non-flammable, and flexible, aligning perfectly with sustainability goals. This winning combination of environmental friendliness and exceptional chemical and physical properties—including high thermal and electrochemical stability—makes PVA-based IL gel electrolytes a tantalizing prospect for researchers developing next-generation, high-performance, and planet-friendly energy storage systems. Following this very strategy, Lee et al. [73] reported fabricating a supercapacitor with a series of imidazolium moiety-based GPE, where EMIM-based IL-GPE showed enhanced electrochemical performance, with the highest capacitance obtained to be 160.6 F/g (at current density 0.5 A/g), energy density 180.67 Wh/kg at a power density 375 W/kg. Jeong et al. [74] designed a full device as a flexible supercapacitor with similar cation moiety with PVA (using reduced graphene oxide (rGO)/single-wall carbon nanotubes (SWCNTs) as electrodes). The cycling stability was analyzed as the function of the bending test; the specific capacitance value of the unbent one was 62.4 F/g whereas with 1000 cycles it retained 89.3% of the initial capacitance. Whereas with 300 degrees bending after 4000 GCD cycles, it retained 64.6% of the initial capacitance value (62.4 F/g at the bent condition). Alexandre et al. [75] used a mixture of poly(1-vinyl-3-propylimidazolium bis(fluorosulfonyl)imide) (poly(VPIFSI)) and EMIMFSI (PIL/IL GPE) in the 50:50 mass ratio, that had shown high ionic conductivity up to 3.7 ms/cm. The electrochemical properties were analyzed in a symmetric supercapacitor with MWCNT electrodes with a maximum operating window of 2.5 V. As shown in **Figure 7a**, the ESR value is 39 Ω, which is on a little higher end compared to the conventional supercapacitors that use ionic liquid directly in a liquid state, but it is lower than that of solid-state cells, this is attributed to the good adhesive properties of PIL/IL GPE to the porous electrode surface. At the low frequencies of the ESR (where the voluminous ions of the IL diffuse directly in the pores of the nanotubes), it shows a straight line close to 90°; this indicates the good capacitive performance of such devices. The symmetric devices, as shown in **Figure 7b,** show nearly triangular shape characteristics of a symmetric double-layer supercapacitor with a specific capacitance of 9.6 F/g, columbic efficiency of 94%, energy density of 8.1 Wh/kg with power density of 537 W/kg, at

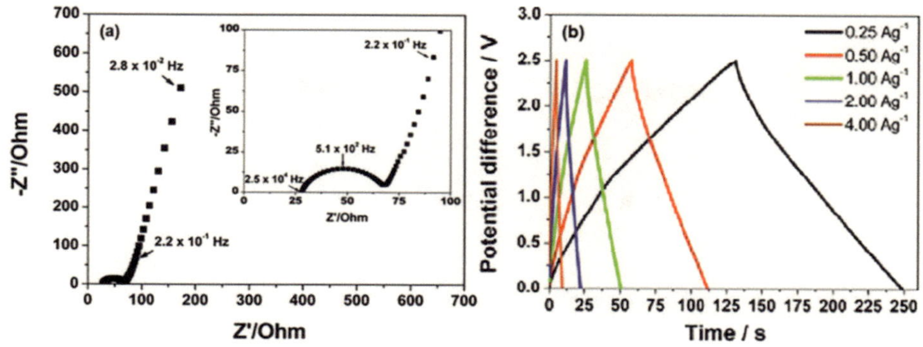

Figure 7.
Electrochemical data obtained for supercapacitor prepared with PIL/IL-GPE (optimized with 50 wt% of IL content) (a) Nyquist plot, (b) galvanostatic charge/discharge curves at the various current densities (0.25 A/g to 4.0 A/g) for potential window 0 to 2.5 V (reprinted/adapted with permission from Ref. [75]).

the current density of 0.5 A/g. The energy efficiency (η = 68.2%) of the PIL/IL GPE has similar values as observed with direct IL [76]-based devices. This higher value of η is considered due to comparatively large size of ionic liquid ions and the semi-solid nature of the system.

Liew et al. [77] introduced the ammonium acetate in PVA to decrease the hardness and glass transition temperature of the PVA backbone and then added BMIMCl to prepare the GPE in the solvent casting method. The EDLC with PE doped with 50% IL gives a maximum intensity of 7.3 mS/cm and from the electrochemical data, it is clear that it also increases the capacitive nature of EDLC. Similarly, Zhang et al. [78] introduced Li_2SO_4 with the same ionic liquid moiety in the PVA backbone to enhance the ionic conductivity as well as the potential window (1.5 V). With the optimized concentration of PVA:BMIMCl:Li_2SO_4 = 1:3:2.2, the maximum ionic conductivity reached is 37 ms/cm even at the bending angle of 180°. The CV data also support the flexibility of the GPE as at bending from 0° to 135°, the capacitance does not show any significant change, indicating good cycling stability. Song et al. proposed the sodium-ion conducting PVA GPE membrane with EMIMTf plasticizer, with NaTf as a sodium salt source. Just by incorporating 10% IL, the optimized GPE (with ionic conductivity of 3.8 mS/cm) shows thermal stability up to 150°C with electrochemical stability up to 4.70 V. This GPE demonstrates remarkable potential for EDLC applications. Electrochemical devices fabricated with this material achieved both 100% columbic efficiency and 100% capacity retention over 1000 cycles. In another study [79], the same EMIMTf/NaTf composition was incorporated in a 2-hydroxyethyl cellulose backbone. Although the ionic conductivity is compromised (maximum 1.11 mS/cm), the electrochemical stability increases further with 4.8 V. The membrane shows a considerable tensile strength and braking strain (4.5 MPa and 94.4%). The highest capacitance value achieved is 204.1 F/g at a current density of 0.4 A/g, with increasing current density (4 A/g), the capacitance value remained up to 120.3 F/g. The Ragone plot reported shows that the highest energy density of 28.2 Wh/kg is obtained at a power density of 280.1 W/kg, whereas the highest power density achieved is 2139.1 W/kg at the energy density of 15.9 Wh/kg. The capacity retention decreases very slowly with increasing cycle number even at moderately high current density of 2A/g. The before and after cycling SEM depicts that the surface of the polymer membrane becomes smoother after cycling; this confirms during charge-discharge

Composition (wt% or vol% IL)	Ionic conductivity (S/cm)	Electrochemical stability window (V)	Specific capacity (F/g)	Power density (kW/kg)	Energy density (Wh/kg)	Long cycling retention (%)	Electrodes	Ref.
PVA:BMImCl:Li2SO4 (1:3:2.2)	37×10^{-3}	1.5	136	3.4	10.6	90% (after 3000 cycles)	Activated carbon	[78]
PVA:EMImTFSI:$LiClO_4$ (7:2:1)	2.8×10^{-3}	3	185	2.1	3.8	88% (after 1000 cycles)	Carbon nanotubes	[80]
PVA:BMImTFSI:LiTFSI (8:1:1)	2.1×10^{-3}	3	200	2.7	5	94% (after 1000 cycles)	Reduced graphene oxide (rGO)	[81]
PVA:$LiClO_4$:EMImTFSI (8:1:1)	1.8×10^{-3}	3.2	165	1.9	3.4	91% (after 1000 cycles)	Nitrogen-doped carbon nanofibers	[82]

Table 1.
Comparison table for PVA-based PIL_GPE.

process, the intimate contact between the membrane and electrode was formed. The electrochemical along with physic-chemical data demonstrates that the GPE may be considered as a good alternative to the conventional energy storage devices (**Table 1**).

PVDF-based IL GPE attracted more attention from researchers due to superior electrochemical and chemical stability, boosting both the power and life span of the energy devices while seamlessly integrating with other materials. For instance, Adusei et al. [83] describe a simple approach to use 3D grapheme-CNT hybrid fibers as electrodes with PVDF-EMIMBF$_4$ ionic gel electrolyte. The device offered a stable potential window of 3.2 V with a high gravimetric energy density of 12.93 Wh/kg with a power density of 1350.25 W/kg (at a current density of 1 A/g). The device also maintained high-capacity retention over 2000 charge-discharge cycles (at a current density of 1A/g). Feng et al. [84] reported the same GPE composition with hierarchical grapheme nanocomposite electrodes. The specific capacitance of the device was calculated even in the bending state, where at the unbent condition, it was showing 180 F/g and 164 F/g (i.e., 91.1%) in the 180° bending condition. Yang et al. [85] reported a stable flexible SC configuration fabricated with PVDF-HFP/tetrathylammonium tetrafluroborate-acetonitrile (Et$_4$NBF$_4$-AN) GPE with carbon nanotube paper@poly(1,5-diaminoanthraquinone)(CNT@PDAA) film electrode (**Figure 8a**). With this flexible configuration, the ionic conductivity has reached 14.4 mS/cm. This high ionic conductivity is attributed to quick high electrolyte uptake; the optimized GPE shows 250% uptake in just 3 minutes. This confirms the perfect interconnectivity offered by the porous connectivity of the cross-linked polymer backbone. The cyclic stability was checked using a continuous charge/discharge study for 15,000 cycles within 0–2.7 V (**Figure 8d**); in the first 2000 cycles, there is a 12% fading capacity, whereas at the end of 15,000 cycles, the fading is just 18%. That indicates the good stability and reversibility of the device. The capacity fading was checked with repeated bending and the negligible capacity fading confirms the excellent flexibility of the device. The ragone plot showed that the device achieves a high-energy density of 41.4 Wh/kg at a power density of 0.41 kW/kg (**Figure 8e**), and holds an energy density of 18.8 Wh/kg at a power density of 15.9 kW/kg. This

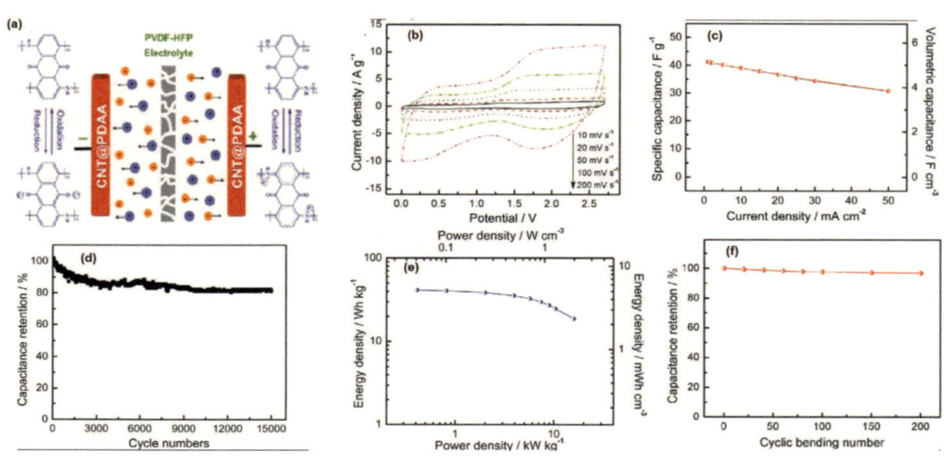

Figure 8.
(a) Schematic representation of charge/discharge process with CNT@PDAA electrodes, (b) CV curves of the optimized electrolyte at different current rates, (c) specific and volumetric capacitance variation with current densities, (d) capacitance retention at unbent condition, (e) Ragone plot, (f) capacitance retention after cyclic bending at angle 145° for the device (reprinted with permission Ref. [85]).

superior capacitance and high-energy density were analyzed to be attributed to the high porosity of the optimized GPE, which is favorable for rapid ion transport, and because of p and n-doping characteristics of PDAA, this contributed to the broad electroactive potential window. Even the nano-sized shell layer of PDAA enhances the utilization of the active materials and facilitates the migration of the electrolyte ions during rapid redox reactions.

In one of the recent studies [86], it has been reported to incorporate 300 wt% IL in PVDF-HFP to attain ionic conductivity of 2.37 mS/cm with 30 wt% sodium thiocynate salt. Due to this large liquid content, it confirms the enhancement of the amorphous region in the polymer matrix that supports the movement of $EMIM^+$ ions throughout the chain. The electrochemical stability window was reported to be 2.1 V with transference number 0.71, which reconfirms that the main component of the GPE has the ionic nature. With a simple activated porous carbon electrode, it offered 18 F/g specific capacitance (calculated from cyclic voltammetry). A similar imidazolium moiety [87] was reported to incorporate in the presence of NH_4Tf salt that shows a high specific capacitance of 165 F/g with a considerably high-energy density of 25.66 Wh/kg with a power density of 14.35 kW/kg. To design such high-performance symmetric EDLC, the electrode composition was changed by the addition of *citrus limetta* pulp-derived AC electrodes (**Table 2**).

Polyethylene oxide (PEO) is also considered a good candidate for GPE because of its excellent mechanical stability, film-forming ability, high ionic conductivity (which comes due to high porosity that provides channels for ion transport), and compatibility with electrode materials. Its polar nature enables ion movement, while its gel formation ensures good contact and mechanical stability. Its wide voltage window and resistance to degradation ensure reliable performance. However, challenges like low-temperature conductivity and narrow electrochemical window limit the application of PEO in practical applications. Liu et al. [82] studied a series of ionic liquid (1-ethyl-3-methylimidazolium tetrafluoroborate) doped (0–70%) solid polymer electrolyte (SPE-PIL(x)) with polypropylene (PP) microporous membrane. Here, the PP membrane was used to enhance thermal as well as mechanical stability. From the DSC curve, it is evident that the melting enthalpy and the corresponding crystallinity decrease with the incorporation of IL into it. With the optimized concentration of 70% IL, the enthalpy drops to 21.6 J/g (with pure PEO ~ 163.7 J/g) and the crystallinity comes to 10.1% (with pure PEO ~ 76.6 J/g). The IL incorporation notably enhances 9.6×10^{-6} to 2.9×10^{-3} S/cm (at RT) as it promotes the salt dissociation in the SPE along with polymer segmental motion through the weakening chain forces. With activated carbon electrodes, the PIL [92] offered a specific capacitance of 158 F/g at a current density of 0.1 A/g, with a high-energy density of 128 Wh/kg with a power density of 302 W/kg. Upon 1000 cycles, the capacity retained up to 90% with a columbic efficiency of 99.37% showing the significantly high stability of the device. The stable morphology and electrochemical stability over a wide temperature window make it suitable for large-scale preparation and predict excellent industrial application potential (**Table 3**).

Singh et al. [48] reported the incorporation of NaI salt (of varying concentrations 1–10 wt%) in the same composition. With the Na ion source, the ionic conductivity of the optimized GPE was obtained to be 4.259×10^{-5} S/cm, which is a little lower than that of the Li counterpart with stable electrochemical window 3.3 V. This may be attributed to the larger size of Na^+ that makes them interact less effectively with the solvent molecules in the electrolyte, hindering their mobility and reducing the overall ionic conductivity. The fabricated EDLC device with activated porous carbon

DOI: http://dx.doi.org/10.5772/intechopen.1005174

Composition (wt% or vol% IL)	Ionic conductivity (S/cm)	Electrochemical stability window (V)	Specific capacity (F/g)	Power density (kW/kg)	Energy density (Wh/kg)	Long cycling retention (%)	Electrodes	Ref.
EMIM-TFSI (20 wt%) in PVDF-HFP	1.02×10^{-2}	3.0	180	1.8	27	92% after 5000 cycles	Carbon aerogel	[88]
EMImBF$_4$ (30 wt%) in PVDF-HFP	1.23×10^{-2}	3.0	210	2.5	38	90% after 1000 cycles	Graphene oxide	[89]
BMIM-PF$_6$ (20 wt%) in PVDF-HFP	2.4×10^{-2}	3.2	165	2.1	33	95% after 1000 cycles	Activated carbon	[90]
EMImTFSI (15 wt%) + LiClO$_4$ (5 wt%) in PVDF-HFP	1.8×10^{-2}	3.5	230	3.2	44	98% after 1000 cycles	Nitrogen-doped carbon nanotubes	[91]

Table 2.
Comparison table for P(VDF-HFP) based PIL_GPE.

Composition (wt% or vol% IL)	Ionic conductivity (S/cm)	Electrochemical stability window (V)	Specific capacity (F/g)	Power density (kW/kg)	Energy density (Wh/kg)	Long cycling retention (%)	Electrodes	Ref.
PEO:LiTFSI:IL (10:1:1)	1.02×10^{-3}	3	200	1.8	3.6	92% (after 5000 cycles)	Activated carbon	[93]
PEO:EMImTFSI (7:3)	1.20×10^{-3}	3.2	180	2.1	4.2	90% (after 1000 cycles)	Carbon nanotubes	[94]
PEO:EMImTFSI:LiTFSI (8:1:1)	2.30×10^{-3}	3.5	250	3.2	5.6	95% (after 1000 cycles)	Reduced graphene oxide (rGO)	[95]
PEO:LiTFSI:EMImTFSI (8:1:1)	1.50×10^{-3}	3	300	2.8	5.4	98% (after 1000 cycles)	N-doped carbon aerogels	[96]
PEO:BMImTFSI (7:3)	1.80×10^{-3}	3	150	1.2	2.4	93% (after 1000 cycles)	Graphene	[97]

Table 3.
Comparison table for PEO-based PIL_GPE.

DOI: http://dx.doi.org/10.5772/intechopen.1005174

Composition (wt% or vol% IL)	Ionic conductivity (S/cm)	Electrochemical stability window (V)	Specific capacity (F/g)	Power density (kW/kg)	Energy density (Wh/kg)	Long cycling retention (%)	Electrodes	Ref.
Chitosan:$LiClO_4$:EMImTFSI (70:15:15)	2.1×10^{-3}	3	220	2.5	4.75	90% (after 1000 cycles)	Activated carbon	[100]
PMMA:PANI:EMImTFSI (70:15:15)	1.7×10^{-3}	2.8	180	1.3	2.34	88% (after 1000 cycles)	Carbon nanotubes	[101]
Cellulose Nanofibril:PIL:EMImTFSI (70:15:15)	2.5×10^{-3}	3.2	240	3	5.76	95% (after 1000 cycles)	Reduced graphene oxide (rGO)	[102]
Silica:PIL:EMImTFSI (60:20:20)	1.9×10^{-3}	3	200	2.2	4	90% (after 1000 cycles)	Activated carbon	[103]

Table 4.
Comparison table for others PIL_GPE.

electrode shows a high specific capacitance of 115 F/g (from CV), energy density of 12.2 Wh/kh, and power density of 2200 W/kg (from GCD), which is promising for a practical EDLC device. A similar study has been reported for Na-based supercapacitor by Manfo et al. [98] using 4 wt% of 1-ethyl-3methyl-imidazolium tricyanomethanide and 5 wt% sodium thiocyanate (NaSCN) as an ion source. The EDLC cell was fabricated using rGO-based symmetric electrodes having potential windows 0–0.8 V, showing the specific capacitance calculated as 0.67 F/g (from CV curves).

The recent years have witnessed explosive growth in the development of ionic liquid (IL)-based gel electrolytes, extending beyond the promising realm of P(VDF-HFP) and PEO-based options. This burgeoning field is rapidly unveiling a tapestry of diverse materials, each boasting unique properties and offering exciting prospects for next-generation energy storage technologies. For instance, Kato et al. [99] reported that ion-conducting nanocomposite films are prepared from poly(2-phenylethyl methacrylate)-grafted cellulose nanocrystals. The optimized GPE with 30 wt% of imidazolium-based ionic liquid shows not only significant enhancement in the toughness (with tensile strength >70 times) but also shows higher ionic conductivity compared to pure PPMA-based electrolyte (**Table 4**).

Mun et al. [104] reported an SSE based on poly(isobornyl methacrylate)-co-poly(ethylene glycol) methyl ether methacrylate (PIBMA-co-PEGMA) (PIBEG)-based comb-like copolymer, that offers a wide potential window and high ionic conductivity. With 85% of EMIMTFSI, the GPE shows ionic conductivity of 4.35 mS/cm due to the fast ion transmission channel of the high polarity of the polymer. This composition shows one of the highest specific capacitances of 48.6 F/g, energy density of 37.12 Wh/kg, and power density of 1174 W/kg in commercial AC symmetric electrodes. Lu et al. [105] reported quasi-solid polymer electrolytes incorporating EMIMTFSI (85 wt%) via a thermally initiated polymerization strategy. The device offered a specific capacitance of 109F/g with an energy density of 38.6 Wh/kg and power density of 130 W/kg (at current density 0.1 mA/cm^2) and it is capable of stable and safe operation with capacitance retention of ~100% with 10,000 cycles. This significant result boosts the application of the in-situ polymerization technique in SC devices.

4. Summary

In the burgeoning field of flexible supercapacitors, replacing traditional liquid electrolytes with gel polymers is revolutionizing supercapacitors, driven by a trifecta of advancements: enhanced safety thanks to leakproof, non-flammable nature; broader operating temperatures that conquer harsh environments; and remarkable flexibility, enabling bendable and stretchable power for wearable electronics. But the benefits go deeper. Tailored ionic conductivity, boosted capacitance through functional groups, and even self-healing capabilities through redox species unlock exciting possibilities. Despite cost considerations and the need for optimized electrode compatibility, in-depth studies using advanced techniques will unveil the secrets of these self-healing gels, paving the way for supercapacitors to break new ground in energy storage and power the future of flexible, safe, and high-performance electronics.

These novel materials hold immense promise for unlocking the full potential of flexible supercapacitors as high-energy storage devices, ultimately paving the way for the widespread adoption of flexible electronics. Advanced in-situ techniques like

X-ray absorption spectroscopy and Fourier-transform infrared spectroscopy will offer crucial insights into these intricate chemical processes, guiding the development of the next generation of high-performance electrolytes and propelling the field of flexible electronics toward unprecedented heights.

Acknowledgements

This work is supported by a CSIR research associateship, funded by CSIR, Government of India (File No: 09/0079(12927)/2021-EMR-I).

Author details

Debalina Deb
The Solid State and Structural Chemistry Unit, Indian Institute of Science, Bangalore, India

*Address all correspondence to: debalinadeb03@gmail.com

References

[1] Olabi AG, Abbas Q, Al Makky A, Abdelkareem MA. Supercapacitors as next generation energy storage devices: Properties and applications. Energy (Oxford, England). 2022;**248**(123617): 123617. DOI: 10.1016/j.energy.2022.123617

[2] Kouchachvili L, Yaïci W, Entchev E. Hybrid battery/supercapacitor energy storage system for the electric vehicles. Journal of Power Sources. 2018;**374**:237-248. DOI: 10.1016/j.jpowsour.2017.11.040

[3] Liu Y, Jiang SP, Shao Z. Intercalation pseudocapacitance in electrochemical energy storage: Recent advances in fundamental understanding and materials development. Materials Today Advances. 2020;**7**:100072. DOI: 10.1016/j.mtadv.2020.100072

[4] Ma Y, Chen D, Fang Z, Zheng Y, Li W, Xu S, et al. High energy density and extremely stable supercapacitors based on carbon aerogels with 100% capacitance retention up to 65,000 cycles. Proceedings of the National Academy of Sciences of the United States of America. 2021;**118**(21). DOI: 10.1073/pnas.2105610118

[5] Yang Z, Tian J, Ye Z, Jin Y, Cui C, Xie Q, et al. High energy and high power density supercapacitor with 3D Al foam-based thick graphene electrode: Fabrication and simulation. Energy Storage Materials. 2020;**33**:18-25. DOI: 10.1016/j.ensm.2020.07.020

[6] Li L, Wu Z, Yuan S, Zhang X-B. Advances and challenges for flexible energy storage and conversion devices and systems. Energy & Environmental Science. 2014;**7**(7):2101. DOI: 10.1039/c4ee00318g

[7] Yang K, Fan Q, Song C, Zhang Y, Sun Y, Jiang W, et al. Enhanced functional properties of porous carbon materials as high-performance electrode materials for supercapacitors. Green Energy and Resources. 2023;**1**(3):100030. DOI: 10.1016/j.gerr.2023.100030

[8] Li Z, Lin J, Li B, Yu C, Wang H, Li Q. Construction of heteroatom-doped and three-dimensional graphene materials for the applications in supercapacitors: A review. Journal of Energy Storage. 2021;**44**(103437):103437. DOI: 10.1016/j.est.2021.103437

[9] Arvas MB, Gürsu H, Gencten M, Sahin Y. Preparation of different heteroatom doped graphene oxide based electrodes by electrochemical method and their supercapacitor applications. Journal of Energy Storage. 2021;**35**(102328):102328. DOI: 10.1016/j.est.2021.102328

[10] Wang R, Li X, Nie Z, Zhao Y, Wang H. Metal/metal oxide nanoparticles-composited porous carbon for high-performance supercapacitors. Journal of Energy Storage. 2021;**38**(102479):102479. DOI: 10.1016/j.est.2021.102479

[11] Dabir MP, Masoudpanah SM, Mamizadeh M. Facile synthesis of free-binder (NiCo)x(CuZn)1-xMoO4 (x = 0, 0.25, 0.5, 0.75, and 1)/nickel foam electrodes for supercapacitor application. Journal of Energy Storage. 2024;**76**(109826):109826. DOI: 10.1016/j.est.2023.109826

[12] Biswal A, Panda PK, Acharya AN, Mohapatra S, Swain N, Tripathy BC, et al. Role of additives in electrochemical deposition of ternary metal oxide microspheres for supercapacitor

applications. ACS Omega. 2020;**5**(7):3405-3417. DOI: 10.1021/acsomega.9b03657

[13] Sharma S, Chand P. Supercapacitor and electrochemical techniques: A brief review. Results in Chemistry. 2023;**5**(100885):100885. DOI: 10.1016/j.rechem.2023.100885

[14] Jalal NI, Ibrahim RI, Oudah MK. A review on supercapacitors: Types and components. Journal of Physics. Conference Series. 2021;**1973**(1):012015. DOI: 10.1088/1742-6596/1973/1/012015

[15] Deb D, Bhattacharya S. Influence of ionic-liquid-tethered Al_2O_3 nanoparticle on the nonisothermal cold crystallization in ionic-liquid-based nanofluids. The Journal of Physical Chemistry. C, Nanomaterials and Interfaces. 2017;**121**(12):6962-6976. DOI: 10.1021/acs.jpcc.6b11845

[16] Bose P, Deb D, Bhattacharya S. Ionic liquid based nanofluid electrolytes with higher lithium salt concentration for high-efficiency, safer, lithium metal batteries. Journal of Power Sources. 2018;**406**:176-184. DOI: 10.1016/j.jpowsour.2018.10.050

[17] Wong CY, Wong WY, Loh KS, Lim KL. Protic ionic liquids as next-generation proton exchange membrane materials: Current status & future perspectives. Reactive & Functional Polymers. 2022;**171**(105160):105160. DOI: 10.1016/j.reactfunctpolym.2022.105160

[18] Khoo KS, Chia WY, Wang K, Chang C-K, Leong HY, Maaris MNB, et al. Development of proton-exchange membrane fuel cell with ionic liquid technology. The Science of the Total Environment. 2021;**793**(148705):148705. DOI: 10.1016/j.scitotenv.2021.148705

[19] Chao L, Niu T, Xia Y, Chen Y, Huang W. Ionic liquid for perovskite solar cells: An emerging solvent engineering technology. Accounts of Materials Research. 2021;**2**(11):1059-1070. DOI: 10.1021/accountsmr.1c00154

[20] Bai S, Da P, Li C, Wang Z, Yuan Z, Fu F, et al. Planar perovskite solar cells with long-term stability using ionic liquid additives. Nature. 2019;**571**(7764):245-250. DOI: 10.1038/s41586-019-1357-2

[21] Jurado LA, Espinosa-Marzal RM. Insight into the electrical double layer of an ionic liquid on graphene. Scientific Reports. 2017;**7**(1):1-12. DOI: 10.1038/s41598-017-04576-x

[22] Han M, Kim H, Leal C, Negrito M, Batteas JD, Espinosa-Marzal RM. Insight into the electrical double layer of ionic liquids revealed through its temporal evolution. Advanced Materials Interfaces. 2020;**7**(24):2001313. DOI: 10.1002/admi.202001313

[23] Jónsson E. Ionic liquids as electrolytes for energy storage applications – A modelling perspective. Energy Storage Materials. 2020;**25**:827-835. DOI: 10.1016/j.ensm.2019.08.030

[24] Watanabe M, Thomas ML, Zhang S, Ueno K, Yasuda T, Dokko K. Application of ionic liquids to energy storage and conversion materials and devices. Chemical Reviews. 2017;**117**(10):7190-7239. DOI: 10.1021/acs.chemrev.6b00504

[25] Romero-Martínez A, Hernández-Guerrero E, Ramírez-Jaramillo E, Martínez-Palou R. Synthesis and evaluation of a pyridinium-based ionic liquid-type surfactant as a new low-dosage methane hydrate inhibitor. Energy & Fuels: An American Chemical Society Journal. 2020;**34**(8):9243-9251. DOI: 10.1021/acs.energyfuels.0c00426

[26] Ab Rahim AH, Yunus NM, Jaffar Z, Allim MF, Othman Zailani NZ, Mohd Fariddudin SA, et al. Synthesis and characterization of ammonium-based protic ionic liquids for carbon dioxide absorption. RSC Advances. 2023;**13**(21):14268-14280. DOI: 10.1039/d3ra01345f

[27] Luo J, Conrad O, Vankelecom IFJ. Physicochemical properties of phosphonium-based and ammonium-based protic ionic liquids. Journal of Materials Chemistry. 2012;**22**(38):20574. DOI: 10.1039/c2jm34359b

[28] Garcia-Quintana L, Jónsson E, Yunis R, MacFarlane DR, Forsyth M, Bond AM, et al. Stabilisation of the superoxide anion in bis(fluorosulfonyl) imide (FSI) ionic liquid by small chain length phosphonium cations: Voltammetric, DFT modelling and spectroscopic perspectives. Electrochemistry Communications. 2021;**127**(107029):107029. DOI: 10.1016/j.elecom.2021.107029

[29] Rennie AJR, Martins VL, Torresi RM, Hall PJ. Ionic liquids containing sulfonium cations as electrolytes for electrochemical double layer capacitors. The Journal of Physical Chemistry. C, Nanomaterials and Interfaces. 2015;**119**(42):23865-23874. DOI: 10.1021/acs.jpcc.5b08241

[30] Trenzado JL, Gutiérrez A, Alcalde R, Atilhan M, Aparicio S. Insights on [BMIM][BF4] and [BMIM][PF6] ionic liquids and their binary mixtures with acetone and acetonitrile. Journal of Molecular Liquids. 2019;**294**(111632):111632. DOI: 10.1016/j.molliq.2019.111632

[31] Ou S, Ren Y, Lu W, Wang J, Yin W. A theoretical investigation of the interaction between small Pd particles and 1-butyl-3-methyl imidazolium ionic liquids with Cl−, BF4− and PF6− anions. Physical Chemistry Chemical Physics: PCCP. 2012;**14**(38):13444. DOI: 10.1039/c2cp41065f

[32] Danten Y, Cabaço MI, Besnard M. Interaction of water highly diluted in 1-alkyl-3-methyl imidazolium ionic liquids with the PF_6^- and BF_4^- anions. The Journal of Physical Chemistry. A. 2009;**113**(12):2873-2889. DOI: 10.1021/jp8108368

[33] Pitawela NR, Shaw SK. Imidazolium triflate ionic liquids' capacitance–potential relationships and transport properties affected by cation chain lengths. ACS Measurement Science Au. 2021;**1**(3):117-130. DOI: 10.1021/acsmeasuresciau.1c00015

[34] Shkrob IA, Marin TW, Zhu Y, Abraham DP. Why bis(fluorosulfonyl) imide is a "magic anion" for electrochemistry. The Journal of Physical Chemistry. C, Nanomaterials and Interfaces. 2014;**118**(34):19661-19671. DOI: 10.1021/jp506567p

[35] Pohlmann S, Olyschläger T, Goodrich P, Alvarez Vicente J, Jacquemin J, Balducci A. Azepanium-based ionic liquids as green electrolytes for high voltage supercapacitors. Journal of Power Sources. 2015;**273**:931-936. DOI: 10.1016/j.jpowsour.2014.09.167

[36] Brandt A, Pohlmann S, Varzi A, Balducci A, Passerini S. Ionic liquids in supercapacitors. MRS Bulletin. 2013;**38**(7):554-559. DOI: 10.1557/mrs.2013.151

[37] Magina S, Barros-Timmons A, Ventura SPM, Evtuguin DV. Evaluating the hazardous impact of ionic liquids – Challenges and opportunities. Journal of Hazardous Materials. 2021;**412**(125215):125215. DOI: 10.1016/j.jhazmat.2021.125215

[38] Lin H, Zhang S, Sun J-K, Antonietti M, Yuan J. Poly(ionic liquid)s with engineered nanopores for energy and environmental applications. Polymer. 2020;**202**(122640):122640. DOI: 10.1016/j.polymer.2020.122640

[39] Lebedeva O, Kultin D, Kustov L. Advanced research and prospects on polymer ionic liquids: Trends, potential and application. Green Chemistry: An International Journal and Green Chemistry Resource: GC. 2023;**25**(22):9001-9019. DOI: 10.1039/d3gc02131a

[40] Durak O, Zeeshan M, Habib N, Gulbalkan HC, Alsuhile AAAM, Caglayan HP, et al. Composites of porous materials with ionic liquids: Synthesis, characterization, applications, and beyond. Microporous and Mesoporous Materials: The Official Journal of the International Zeolite Association. 2022;**332**(111703):111703. DOI: 10.1016/j.micromeso.2022.111703

[41] Taghavikish M, Subianto S, Gu Y, Sun X, Zhao XS, Choudhury NR. A poly(ionic liquid) gel electrolyte for efficient all solid electrochemical double-layer capacitor. Scientific Reports. 2018;**8**(1):1-10. DOI: 10.1038/s41598-018-29028-y

[42] Mendhe A, Panda HS. A review on electrolytes for supercapacitor device. Discover Materials. 2023;**3**(1). DOI: 10.1007/s43939-023-00065-3

[43] Huo S, Sheng L, Xue W, Wang L, Xu H, Zhang H, et al. Challenges of polymer electrolyte with wide electrochemical window for high energy solid-state lithium batteries. InfoMat. 2023;**5**(3):e12394. DOI: 10.1002/inf2.12394

[44] Aruchamy K, Ramasundaram S, Divya S, Chandran M, Yun K, Oh TH. Gel polymer electrolytes: Advancing solid-state batteries for high-performance applications. Gels (Basel, Switzerland). 2023;**9**(7):585. DOI: 10.3390/gels9070585

[45] Boz B, Dev T, Salvadori A, Schaefer JL. Review—Electrolyte and electrode designs for enhanced ion transport properties to enable high performance lithium batteries. Journal of the Electrochemical Society. 2021;**168**(9):090501. DOI: 10.1149/1945-7111/ac1cc3

[46] Adedoja OS, Sadiku ER, Hamam Y. An overview of the emerging technologies and composite materials for supercapacitors in energy storage applications. Polymers. 2023;**15**(10):2272. DOI: 10.3390/polym15102272

[47] Wang Y, Xue K, Zhang X, Zhang X, Ma P, Yang B, et al. High-voltage electrochemical double layer capacitors enabled by polymeric ionic liquid. Electrochimica Acta. 2023;**441**(141829):141829. DOI: 10.1016/j.electacta.2023.141829

[48] Singh A, Dhapola PS, Konwar S, Sharma T, Jun HK, Singh D, et al. Ionic liquid (1-ethyl-3-methyltricynomethanide) doped polymer electrolyte (polyvinyl alcohol) for sustainable energy devices. Journal of Science Advanced Materials and Devices. 2023;**8**(3):100566. DOI: 10.1016/j.jsamd.2023.100566

[49] Zhou D, Shanmukaraj D, Tkacheva A, Armand M, Wang G. Polymer electrolytes for lithium-based batteries: Advances and prospects. Chem. 2019;**5**(9):2326-2352. DOI: 10.1016/j.chempr.2019.05.009

[50] Chattopadhyay J, Pathak TS, Santos DMF. Share an Online Entry "Applications of Polymer Electrolytes in Encyclopedia". Pub; Diogo M. F.

Santos. 2023. Available from: https://encyclopedia.pub/entry/50076

[51] Ye Y-S, Rick J, Hwang B-J. Ionic liquid polymer electrolytes. Journal of Materials Chemistry. A, Materials for Energy and Sustainability. 2013;**1**(8):2719-2743. DOI: 10.1039/c2ta00126h

[52] Pan S, Yao M, Zhang J, Li B, Xing C, Song X, et al. Recognition of ionic liquids as high-voltage electrolytes for supercapacitors. Frontiers in Chemistry. 2020;**8**:261. DOI: 10.3389/fchem.2020.00261

[53] Yu X-Y, Xiao M, Wang S-J, Zhao Q-Q, Meng Y-Z. Fabrication and characterization of PEO/PPC polymer electrolyte for lithium-ion battery. Journal of Applied Polymer Science. 2010;**115**(5):2718-2722. DOI: 10.1002/app.29915

[54] Leones R, Sabadini RC, Sentanin FC, Esperança JMSS, Pawlicka A, Silva MM. Polymer electrolytes for electrochromic devices through solvent casting and sol-gel routes. Solar Energy Materials and Solar Cells: An International Journal Devoted to Photovoltaic, Photothermal, and Photochemical Solar Energy Conversion. 2017;**169**:98-106. DOI: 10.1016/j.solmat.2017.04.047

[55] Liu B, Huang Y, Zhao L, Huang Y, Song A, Lin Y, et al. A novel non-woven fabric supported gel polymer electrolyte based on poly(methylmethacrylate-polyhedral oligomeric silsesquioxane) by phase inversion method for lithium ion batteries. Journal of Membrane Science. 2018;**564**:62-72. DOI: 10.1016/j.memsci.2018.07.014

[56] Liu B, Huang Y, Cao H, Zhao L, Huang Y, Song A, et al. A novel polyacrylonitrile-based porous structure gel polymer electrolyte composited by incorporating polyhedral oligomeric silsesquioxane by phase inversion method. Journal of Solid State Electrochemistry: Current Research and Development in Science and Technology. 2018;**22**(6):1771-1783. DOI: 10.1007/s10008-017-3877-8

[57] Li W, Wu Y, Wang J, Huang D, Chen L, Yang G. Hybrid gel polymer electrolyte fabricated by electrospinning technology for polymer lithium-ion battery. European Polymer Journal. 2015;**67**:365-372. DOI: 10.1016/j.eurpolymj.2015.04.014

[58] Nikodimos Y, Su W-N, Shitaw KN, Jiang S-K, Abrha LH, Weret MA, et al. Multifunctional electrospun PVDF-HFP gel polymer electrolyte membrane suppresses dendrite growth in anode-free li metal battery. Energy Storage Materials. 2023;**61**(102861):102861. DOI: 10.1016/j.ensm.2023.102861

[59] Tan H, Jia S. Electrospun polyacrylonitrile/polyvinylidene fluoride/boehmite separator and gel polymer electrolyte polyethylene oxide/polyvinylidene fluoride-hexafluoropropylene/lithium bis(trifluoromethanesulfonyl)imide/boehmite composite separator are used for fast charging. Macromolecular Materials and Engineering. 2023;**308**:2300071. DOI: 10.1002/mame.202300071

[60] Sun M, Zeng Z, Zhong W, Han Z, Peng L, Cheng S, et al. *In-situ* polymerization methods for polymer-based solid-state lithium batteries. Batteries & Supercaps. 2022;**5**:e202200338. DOI: 10.1002/batt.202200338

[61] Liu Q, Wang L, He X. Toward practical solid-state polymer lithium batteries by in situ polymerization process: A review. Advanced Energy Materials. 2023;**13**:2300798. DOI: 10.1002/aenm.202300798

[62] Mizuta Y, Kobiro K, Nishiwaki N, Sugimoto R. Graft polymerization of methyl methacrylate on the surface of poly(vinylidene fluoride) using tributylborane as an initiator. Polymer Journal. 2023;**56**(2):79-85. DOI: 10.1038/s41428-023-00853-x

[63] Wu F, Zhang K, Liu Y, Gao H, Bai Y, Wang X, et al. Polymer electrolytes and interfaces toward solid-state batteries: Recent advances and prospects. Energy Storage Materials. 2020;**33**:26-54. DOI: 10.1016/j.ensm.2020.08.002

[64] Kim S, Ryu J, Rim J, Hong D, Kang J, Park S. Vinyl-integrated in situ cross-linked composite gel electrolytes for stable lithium metal anodes. ACS Applied Energy Materials. 2021;**4**(3):2922-2931. DOI: 10.1021/acsaem.1c00327

[65] Liu M, Wang Y, Li M, Li G, Li B, Zhang S, et al. A new composite gel polymer electrolyte based on matrix of PEGDA with high ionic conductivity for lithium-ion batteries. Electrochimica Acta. 2020;**354**(136622):136622. DOI: 10.1016/j.electacta.2020.136622

[66] Choi K-H, Kim S-H, Ha H-J, Kil E-H, Lee CK, Lee SB, et al. Compliant polymer network-mediated fabrication of a bendable plastic crystal polymer electrolyte for flexible lithium-ion batteries. Journal of Materials Chemistry. A, Materials for Energy and Sustainability. 2013;**1**(17):5224. DOI: 10.1039/c3ta10368d

[67] Zhou D, He Y-B, Liu R, Liu M, Du H, Li B, et al. In situ synthesis of a hierarchical all-solid-state electrolyte based on nitrile materials for high-performance lithium-ion batteries. Advanced Energy Materials. 2015;**5**:1500353. DOI: 10.1002/aenm.201500353

[68] Chen M, Zhong M, Johnson JA. Light-controlled radical polymerization: Mechanisms, methods, and applications. Chemical Reviews. 2016;**116**(17):10167-10211. DOI: 10.1021/acs.chemrev.5b00671

[69] Siccardi S, Amici J, Colombi S, Carvalho JT, Versaci D, Quartarone E, et al. UV-cured self-healing gel polymer electrolyte toward safer room temperature lithium metal batteries. Electrochimica Acta. 2022;**433**:141265. DOI: 10.1016/j.electacta.2022.141265

[70] Niu H, Ding M, Zhang N, Guo X, Guan P, Hu X. Ionic liquid-modified silicon nanoparticles composite gel polymer electrolyte for high-performance lithium batteries. ChemElectroChem. 2023;**10**:e202201015. DOI: 10.1002/celc.202201015

[71] Ha H-J, Kwon YH, Kim JY, Lee S-Y. A self-standing, UV-cured polymer networks-reinforced plastic crystal composite electrolyte for a lithium-ion battery. Electrochimica Acta. 2011;**57**:40-45. DOI: 10.1016/j.electacta.2011.03.101

[72] Li J, Tang J, Yuan J, Zhang K, Shao Q, Sun Y, et al. Interactions between graphene and ionic liquid electrolyte in supercapacitors. Electrochimica Acta. 2016;**197**:84-91. DOI: 10.1016/j.electacta.2016.03.036

[73] Lee KS, Jeong HT. Development and optimization of ionic liquid based gel polymer electrolyte for all solid-state supercapacitor. Journal of Energy Storage. 2021;**42**(103001):103001. DOI: 10.1016/j.est.2021.103001

[74] Jeong HT, Du JF, Kim YR. Development of flexible energy storage device by using polymer electrolyte based on ionic liquid. ChemistrySelect. 2017;**2**(21):6057-6061. DOI: 10.1002/slct.201701073

[75] Alexandre SA, Silva GG, Santamaría R, Trigueiro JPC, Lavall RL.

A highly adhesive PIL/IL gel polymer electrolyte for use in flexible solid state supercapacitors. Electrochimica Acta. 2019;**299**:789-799. DOI: 10.1016/j.electacta.2019.01.029

[76] Ortega PFR, González Z, Blanco C, Silva GG, Lavall RL, Santamaría R. Biliquid supercapacitors: A simple and new strategy to enhance energy density in asymmetric/hybrid devices. Electrochimica Acta. 2017;**254**:384-392. DOI: 10.1016/j.electacta.2017.09.095

[77] Liew C-W, Ramesh S, Arof AK. Good prospect of ionic liquid based-poly(vinyl alcohol) polymer electrolytes for supercapacitors with excellent electrical, electrochemical and thermal properties. International Journal of Hydrogen Energy. 2014;**39**(6):2953-2963. DOI: 10.1016/j.ijhydene.2013.06.061

[78] Zhang X, Wang L, Peng J, Cao P, Cai X, Li J, et al. A flexible ionic liquid gelled PVA-Li_2SO_4 polymer electrolyte for semi-solid-state supercapacitors. Advanced Materials Interfaces. 2015;**2**(15). DOI: 10.1002/admi.201500267

[79] Liu J, Ye Z, Hu X, Ahmed S, Song S. High-performance Na-ion conducting polymer gel membrane for supercapacitor applications. ACS Applied Polymer Materials. 2022;**4**(1):280-288. DOI: 10.1021/acsapm.1c01267

[80] Chodankar NR, Dubal DP, Lokhande AC, Lokhande CD. Ionically conducting PVA–$LiClO_4$ gel electrolyte for high performance flexible solid state supercapacitors. Journal of Colloid and Interface Science. 2015;**460**:370-376. DOI: 10.1016/j.jcis.2015.08.046

[81] Alipoori S, Mazinani S, Aboutalebi SH, Sharif F. Review of PVA-based gel polymer electrolytes in flexible solid-state supercapacitors: Opportunities and challenges. Journal of Energy Storage. 2020;**27**(101072):101072. DOI: 10.1016/j.est.2019.101072

[82] Peng S, Jiang X, Xiang X, Chen K, Chen G, Jiang X, et al. High-performance and flexible solid-state supercapacitors based on high toughness and thermoplastic poly(vinyl alcohol)/NaCl/glycerol supramolecular gel polymer electrolyte. Electrochimica Acta. 2019;**324**(134874):134874. DOI: 10.1016/j.electacta.2019.134874

[83] Adusei PK, Kanakaraj SN, Gbordzoe S, Johnson K, DeArmond D, Hsieh Y-Y, et al. A scalable nano-engineering method to synthesize 3D-graphene-carbon nanotube hybrid fibers for supercapacitor applications. Electrochimica Acta. 2019;**312**:411-423. DOI: 10.1016/j.electacta.2019.04.179

[84] Feng L, Wang K, Zhang X, Sun X, Li C, Ge X, et al. Flexible solid-state supercapacitors with enhanced performance from hierarchically graphene nanocomposite electrodes and ionic liquid incorporated gel polymer electrolyte. Advanced Functional Materials. 2018;**28**:1704463. DOI: 10.1002/adfm.201704463

[85] Yang C, Sun M, Wang X, Wang G. A novel flexible supercapacitor based on cross-linked PVDF-HFP porous organogel electrolyte and carbon nanotube paper@π-conjugated polymer film electrodes. ACS Sustainable Chemistry & Engineering. 2015;**3**(9):2067-2076. DOI: 10.1021/acssuschemeng.5b00334

[86] Dhawan R, Singh A, Kumar S, Dhapola PS, Alheety MA, Yahya MZA, et al. Futuristic approach towards replacement of aqueous electrolyte with solid polymer electrolyte for supercapacitor applications. Journal of Electronic Materials.

2023;**52**(7):4295-4301. DOI: 10.1007/s11664-023-10212-4

[87] Singh K, Kumar R, Kaur A. Novel hierarchical porous carbon derived from biomass Citrus limetta pulp for high-performance quasi-solid-state supercapacitor electrodes. Journal of Energy Storage. 2023;**71**(108121):108121. DOI: 10.1016/j.est.2023.108121

[88] Tuhania P, Singh PK, Bhattacharya B, Dhapola PS, Yadav S, Shukla PK, et al. PVDF-HFP and 1-ethyl-3-methylimidazolium thiocyanate–doped polymer electrolyte for efficient supercapacitors. High Performance Polymers. 2018;**30**(8):911-917. DOI: 10.1177/0954008318772009

[89] Sarno M, Scudieri C, Ponticorvo E, Baldino L, Cardea S, Reverchon E. High performance PVDF HFP_RuO2 supercapacitors production by supercritical drying. The Journal of Supercritical Fluids. 2021;**176**(105323):105323. DOI: 10.1016/j.supflu.2021.105323

[90] Ciurduc DE, Boaretto N, Fernández-Blázquez JP, Marcilla R. Development of high performing polymer electrolytes based on superconcentrated solutions. Journal of Power Sources. 2021;**506**(230220):230220. DOI: 10.1016/j.jpowsour.2021.230220

[91] Kang DA, Kim K, Karade SS, Kim H, Hak Kim J. High-performance solid-state bendable supercapacitors based on PEGBEM-g-PAEMA graft copolymer electrolyte. Chemical Engineering Journal (Lausanne, Switzerland: 1996). 2020;**384**:123308. DOI: 10.1016/j.cej.2019.123308

[92] Liu W, Li Z, Pan F, He Q, Zhang Q. Solid polymer electrolytes reinforced with porous polypropylene separators for all-solid-state supercapacitors. RSC Advances. 2023;**13**(49):34652-34659. DOI: 10.1039/d3ra05899a

[93] Asfaw HD, Kucernak A, Greenhalgh ES, Shaffer MSP. Electrochemical performance of supercapacitor electrodes based on carbon aerogel-reinforced spread tow carbon fiber fabrics. Composites Science and Technology. 2023;**238**(110042):110042. DOI: 10.1016/j.compscitech.2023.110042

[94] Chen H, Adekoya D, Hencz L, Ma J, Chen S, Yan C, et al. Stable seamless interfaces and rapid ionic conductivity of Ca–CeO_2/LiTFSI/PEO composite electrolyte for high-rate and high-voltage all-solid-state battery. Advanced Energy Materials. 2020;**10**:2000049. DOI: 10.1002/aenm.202000049

[95] Bello UY, Kumar S, Singh PK, Kakroo S, Madkhli AY, Singh S, et al. Insight into the use of ionic liquid-based polymer electrolyte for super capacitor application. High Performance Polymers. 2023;**35**(1):36-55. DOI: 10.1177/09540083221111030

[96] Zhai Z, Ren B, Zheng Y, Xu Y, Wang S, Zhang L, et al. Nitrogen-containing carbon aerogels with high specific surface area for supercapacitors. ChemElectroChem. 2019;**6**(24):5993-6001. DOI: 10.1002/celc.201901951

[97] Jamil R, Silvester DS. Ionic liquid gel polymer electrolytes for flexible supercapacitors: Challenges and prospects. Current Opinion in Electrochemistry. 2022;**35**(101046):101046. DOI: 10.1016/j.coelec.2022.101046

[98] Manfo TA, Konwar S, Singh PK, Mehra RM, Kumar Y, Gupta M. PEO + NaSCN and ionic liquid based polymer electrolyte for supercapacitor. Materials Today: Proceedings. 2021;**34**:802-812. DOI: 10.1016/j.matpr.2020.05.340

[99] Kato R, Lettow JH, Patel SN, Rowan SJ. Ion-conducting thermoresponsive films based on polymer-grafted cellulose nanocrystals. ACS Applied Materials & Interfaces. 2020;**12**(48):54083-54093. DOI: 10.1021/acsami.0c16059

[100] Yamagata M, Soeda K, Ikebe S, Yamazaki S, Ishikawa M. Chitosan-based gel electrolyte containing an ionic liquid for high-performance nonaqueous supercapacitors. Electrochimica Acta. 2013;**100**:275-280. DOI: 10.1016/j.electacta.2012.05.073

[101] Tamilarasan P, Ramaprabhu S. Stretchable supercapacitors based on highly stretchable ionic liquid incorporated polymer electrolyte. Materials Chemistry and Physics. 2014;**148**(1-2):48-56. DOI: 10.1016/j.matchemphys.2014.07.010

[102] Pietrzyk-Thel P, Jain A, Bochenek K, Michalska M, Basista M, Szabo T, et al. Flexible, tough and high-performing ionogels for supercapacitor application. Journal of Materiomics. 2024. DOI: 10.1016/j.jmat.2024.01.008

[103] Kowsari E. High-performance supercapacitors based on ionic liquids and a graphene nanostructure. In: Handy S, editor. Ionic Liquids – Current State of the Art. London, UK: InTech; 2015

[104] Mun WJ, Kim B, Moon SJ, Kim JH. Multifunctional, bicontinuous, flexible comb copolymer electrolyte for solid-state supercapacitors. Chemical Engineering Journal (Lausanne, Switzerland: 1996). 2023;**454**:140386. DOI: 10.1016/j.cej.2022.140386

[105] Lu H, Wang P, Ma Y, Liu M, Chang L, Feng R, et al. *In situ*-fabricated quasi-solid polymer electrolytes incorporating an ionic liquid for flexible supercapacitors. Sustainable Energy & Fuels. 2024;**8**(2):358-368. DOI: 10.1039/d3se01171b

Chapter 6

Isopiestic Method-Thermodynamic Method to Determine the Activity Coefficients of Components of a Solution

Parisa Ghasemi Ilkhechi

Abstract

The isopiestic method is a thermodynamic technique used to measure the activity coefficients of components in a solution. In the context of ionic liquids, which are salts in liquid state below 100°C, the isopiestic method can be applied to study their behavior and properties in various solutions. In the isopiestic method, two solutions with different concentrations of a solute are prepared, and the vapor pressure of the solutions is measured at the same temperature. By comparing the vapor pressures, the activity coefficients of the solute in the solutions can be determined. When it comes to ionic liquids, this method can be used to explore their interactions with other substances, such as different solvents or salts. Understanding these interactions is crucial for various applications of ionic liquids, including their use as green solvents, electrolytes in batteries, and in other chemical processes.

Keywords: isopiestic method, activity coefficients, ionic liquids, electrolytes, solvents

1. Introduction

The isotopic method in ionic liquids involves using isotopically labeled compounds to trace and understand various processes within the liquid. This can include studying diffusion, structure, and interactions. Isotopic labeling allows researchers to track specific atoms or molecules and gain insights into the behavior and mechanisms of ionic liquids at a molecular level. It is a powerful technique for elucidating the properties and dynamics of these unique liquid systems.

Achieving the thermodynamic equilibrium point between two solutions is the primary requirement for this approach. The solutions' initial compositions alter and eventually find equilibrium when they come into contact with one another during the vapor phase and exchange solvents. The solvents' chemical potential, activity, and hence their vapor pressure are all equivalent when the solutions are in equilibrium. An example of an isopistic device is shown in **Figure 1**.

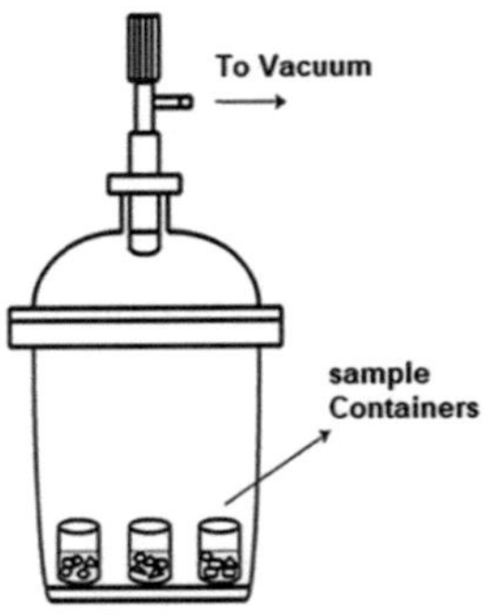

Figure 1.
Schematic of isopistic device.

The heat of evaporation from the solvent in the solutions' evaporation and distillation causes the solutions' temperatures to differ from one another during the solvent exchange process. The entire system is housed in a thermostat environment in order to avoid this circumstance and isotherm of the solutions. We weigh every solution both before and after equilibrium. The method's correctness is contingent upon the precision of the analytical measurements and the reporting of the reference solution's vapor pressure data. At the point when the temperature and strain of all stages become equivalent, thermodynamic harmony is laid out among them and the action of the dissolvable is equivalent in all stages on the grounds that the substance capability of the dissolvable in every one of the arrangements in a shut framework has a similar worth:

$$a_1^{\alpha} = a_2^{\beta} \tag{1}$$

The accompanying relationship exists between dissolvable movement and osmotic coefficient:

$$Lna_1 = -v_s.m_s.M_1.\varnothing_s \tag{2}$$

where M_1 is the molecular mass of the dissolvable (kg/mol), m_s is the molality of the non-unstable solute (mol/kg), $\varnothing_s$ is the osmotic coefficient of the arrangement, and sν is the amount of the stoichiometric coefficients [1–3].

As indicated by relations (1) and (2), we will have:

$$\begin{aligned} &v_{\alpha}.\varnothing_{\alpha}.m_{\alpha} = v_{\beta}.\varnothing_{\beta}.m_{\beta} \\ &\varnothing_{\beta} = \left(v_{\alpha}.m_{\alpha} / v_{\beta}.m_{\beta}\right).\varnothing_{\alpha} \quad \Rightarrow \varnothing_{\beta} = R\varnothing_{\alpha} \end{aligned} \tag{3}$$

When the values of R and Ln $\gamma_{\pm A}$ are available, the value of Ln $\gamma_{\pm B}$ can be calculated as follows:

$$Ln\gamma_{\pm B} = Ln\gamma_{\pm A} + \int_0^{m_A} dLn\frac{m_A}{m_B} + \int_0^{m_A} (R-1)dLn\gamma_A m_A \tag{4}$$

DOI: http://dx.doi.org/10.5772/intechopen.1004797

2. The application of ionic liquids in other electrochemical methods

Ionic liquids, besides isopiestic method, are also utilized in other electrochemical methods such as electrode composition, current composition, electrolyte composition, and more. The unique characteristic of ionic liquids includes their high ionic conductivity and wide electrochemical windows, indicating their high capability for electrochemical operations. This capability allows us to adjust the properties of ionic liquids according to various needs.

One common electrochemical application of ionic liquids is in electrolysis processes, where they can serve as electrolytes due to their high ionic conductivity. Additionally, ionic liquids are used in electrochemical sensors and biosensors due to their ability to solubilize a wide range of analytes and their compatibility with various electrode materials. Furthermore, ionic liquids are employed in electrodeposition and electrosynthesis reactions, where they can act as solvents, electrolytes, or both. Their wide electrochemical stability window allows for the deposition of metals and the synthesis of various organic compounds under mild conditions. Overall, the unique properties of ionic liquids make them valuable components in various electrochemical methods beyond the isopiestic method.

3. The concentration range of ionic liquids used in the isopiestic method for determining activity coefficients

In the isopiestic method, solutions with low concentrations are typically used to determine the activity coefficients of solutes. This is because:

1. At low concentrations, the assumptions of ideal behavior for solutions are usually valid, leading to more accurate thermodynamic calculations.

2. Measurement of vapor pressures of solutions and water vapor (as a reference) is easily performed and provides higher accuracy at low concentrations.

Therefore, in the isopiestic method, solutions with low concentrations, particularly in the range of 0.01–0.1 molar, are commonly employed [4].

4. What are ionic liquids and what are their uses in chemistry?

Ionic liquids can be employed in the isopiestic method to measure their activity coefficients in various solutions. For example, researchers might investigate the behavior of an ionic liquid in water or in a mixed solvent system using the isopiestic method.

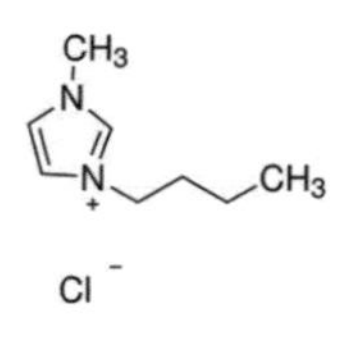

Figure 2.
Composition structure 1-butyl-3-methylimidazolium chloride.

CAS Nr	Sum formula	Molecular weight	Melting point	Density	Solubility
[79917–90-1]	C8H15ClN	174.67	65°C	1.08 Kg/m^3	DMSO (Slightly), Methanol (Sparingly)

Table 1.
Physical and chemical properties of [BMIM]Cl.

Let us consider an ionic liquid like 1-butyl-3-methylimidazolium chloride ([BMIM]Cl). **Figure 2** shows the structure of [BMIM]Cl. To determine its activity coefficients, researchers could create solutions of [BMIM]Cl in different concentrations and use the isopiestic method to measure the vapor pressures of these solutions. By maintaining the same water activity in these solutions and comparing their vapor pressures, scientists can calculate the activity coefficients of [BMIM]Cl in the respective solutions [5].

This information is valuable for understanding how [BMIM]Cl behaves in different environments and can be crucial for various applications, such as in electrolyte systems, catalysis, or solvent-related processes. Physical and chemical properties of [BMIM]Cl are given in **Table 1**.

5. What are ionic liquids and what are their uses in chemistry?

Ionic liquids are salts that are in a liquid state at relatively low temperatures, often below the boiling point of water. Unlike traditional salts, which have high melting points, ionic liquids are composed entirely of ions (charged particles) and exhibit unique properties due to their ionic nature. Typically, they consist of large organic cations paired with various anions. The combination of different cations and anions allows for a wide range of ionic liquids with diverse properties.

5.1 Some of ionic liquids used in chemistry

1. Solvents: Ionic liquids are excellent solvents for a wide range of compounds, including gases, metals, and organic molecules. They can dissolve a variety of substances that are insoluble in traditional organic solvents, making them valuable in chemical synthesis and catalysis [6].

2. Green Chemistry: Ionic liquids are often considered more environmentally friendly than traditional organic solvents. They are non-volatile and can be recycled, reducing environmental pollution. Their use in green chemistry initiatives has gained attention for sustainable chemical processes.

3. Catalysis: Ionic liquids can serve as catalysts or co-catalysts in various chemical reactions. They can stabilize reactive intermediates and enhance reaction rates, making them useful in both academic research and industrial applications.

4. Electrolytes: Ionic liquids are used in batteries and supercapacitors as electrolytes. Their high ionic conductivity and stability under a wide range of conditions make them attractive for energy storage devices. Some examples of ionic liquids used in batteries are given in **Table 2**.

([C5H12NO][N(SO2F)2])
([C8H17NO][N(CF3SO2)2]
([C6H11N2][N(CN)2])

Table 2.
Some examples of ionic liquids used in batteries.

5. Extraction and Separation: Ionic liquids are employed in extraction and separation processes due to their selectivity for specific compounds. They can extract valuable chemicals from natural sources or separate different components of a mixture efficiently.

6. Catalyst Immobilization: Ionic liquids can serve as a medium for immobilizing catalysts. Immobilized catalysts are easier to handle, recover, and reuse, making the overall catalytic process more efficient and sustainable.

7. Gas Absorption: Some ionic liquids have the ability to absorb gases like carbon dioxide and sulfur dioxide. This property is essential in gas purification and separation processes, especially in environmental and industrial applications. **Figure 3** shows the processes of gas purification and separation by ionic liquids.

8. Lubricants: Ionic liquids can be used as lubricants, especially in situations where traditional lubricants fail, such as at extreme temperatures or in vacuum environments.

9. Nanotechnology: Ionic liquids are utilized as suitable solvents for the synthesis of nanoparticles and nanomaterials. Some examples of nanoparticle synthesis using ionic liquids are given in **Table 3**. The steps of nanoparticle synthesis are shown in **Figure 4**.

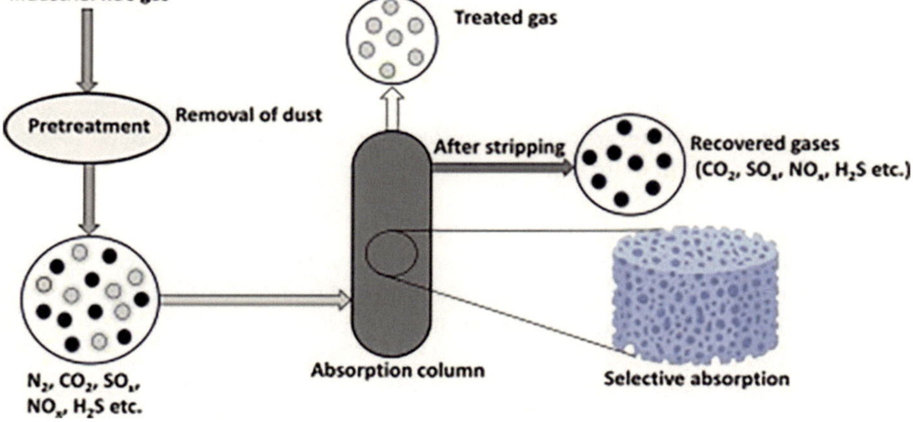

Figure 3.
Processes of gas purification and separation by ionic liquids.

Gold Nanoparticles
Silver Nanoparticles
Magnetic Nanoparticles
Metal Oxide Nanoparticles
Carbon Nanomaterials

Table 3.
Some examples of nanoparticle synthesis.

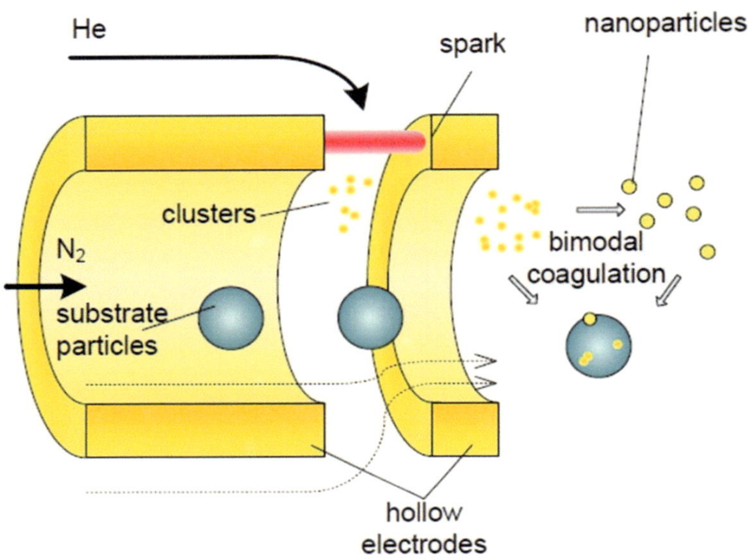

Figure 4.
The steps of nanoparticle synthesis.

6. Obtaining the activity coefficient for ionic solutions using the isopiestic method?

The isopiestic method is a technique used to measure the activity coefficients of solutes in solutions. In the context of ionic liquids, which are salts that are in a liquid state at relatively low temperatures, the isopiestic method can be employed to determine their activity coefficients [7].

Here is how it works:

1. Preparation of Solutions: Prepare two solutions with different concentrations of the ionic liquid. These solutions should have the same water activity, which means they have the same ratio of vapor pressures.

2. Equilibration: Allow the solutions to equilibrate in a closed system until the water activity in both solutions is the same. This is usually achieved by keeping the solutions in a controlled environment until no further changes in water activity are observed.

3. Measurement: Measure the vapor pressures of both solutions once equilibrium is reached. The vapor pressure is related to the activity coefficients of the components in the solutions.

4. Calculation: By comparing the vapor pressures and knowing the solution concentrations, you can calculate the activity coefficients of the ionic liquid in each solution.

5. Analysis: Analyze the data to obtain the activity coefficients at different concentrations of the ionic liquid. This information is valuable for understanding the behavior of ionic liquids in various applications, such as in electrolytes, solvents, or green chemistry processes.

7. How activity coefficients of ionic liquids can help us determine their usefulness for various applications such as solvents, catalysis, electrolytes, extraction and separation, catalyst immobilization, gas adsorption, and lubricants?

The activity coefficients of ionic liquids, including their physical and chemical properties such as density, viscosity, ion exchange, gas solubility, and solubility, can play a crucial role in determining their usefulness for various applications. For instance, for solubility applications, the high and tunable activity coefficients of ionic liquids can be used as alternative solvents to traditional solvents like water or organic oils. Their high activity coefficients and tunability are among the advantages of this application.

Moreover, for catalysis, the activity coefficients of ionic liquids can be utilized to stabilize catalysts and enhance their activity in chemical reactions. Given the diverse properties of ionic liquids and the variety in their activity coefficients, these substances can be used for various applications including extraction, separation, gas absorption, and lubricants, and activity coefficients can play an important role in these areas [8].

8. Conclusion

The isopiestic method in ionic liquids involves using isotopically labeled compounds to trace and understand various processes within the liquid. This can include studying diffusion, structure, and interactions. Isotopic labeling allows researchers to track specific atoms or molecules and gain insights into the behavior and mechanisms of ionic liquids at a molecular level. It is a powerful technique for elucidating the properties and dynamics of these unique liquid systems. The concentration range of ionic liquids used in the isopiestic method for determining activity coefficients typically falls within the lower concentration range. This range is generally around 0.01–0.1 molar or even lower, as within this range, the ideal behavior assumptions of the solutions can be better satisfied.

Acknowledgements

I am grateful to “IntechOpen”, the world’s leading publisher of Open Access books.

I would also like to express my gratitude to "Prof. Pradip K. Bhowmik", the academic editor, and "Dorian Salatic", the publishing process manager who helped me in completing this chapter of the book Ionic liquids.

Appendices and nomenclature

a. The Gibbs energy of the above system $(M_1X_1\text{–}M_2X_2\text{–}H_2O)$ consists of two terms, viz.

$$G = G^{DH} + G^{Hy}$$

b. where G^{DH} is the Debye Hockel contribution and G^{Hy} is the pseudo-ideal mixture description of the resulting species resulting from the mole fraction of component X.

For the ternary aqueous solution of electrolyte A + electrolyte B under isopistic equilibrium, a linear empirical relationship was proposed using Zdanowski [9].

$$\frac{m_A}{m_A^{\circ}} + \frac{m_B}{m_B^{\circ}} = 1\left(a_W = constant\right)$$

Author details

Parisa Ghasemi Ilkhechi
Faculty of Basic Sciences, Physical Chemistry, Department of Chemistry, Azarbaijan Shahid Madani University, Tabriz-Ilkhechi, Iran

*Address all correspondence to: pgh.chemist@gmail.com

References

[1] Clegg SL, Pitzer KS. Thermodynamics of multicomponent, miscible, ionic solutions: Generalized equations for symmetrical electrolytes. The Journal of Physical Chemistry. 1992;**96**(8):3513-3520

[2] Pitzer KS. Ion interaction approach: Theory and data correlation. Activity coefficients in electrolyte solutions. 1991;**2**:75-153

[3] Jaretun A, Aly G. New local composition model for electrolyte solutions: Multicomponent systems. Fluid Phase Equilibria. 2000;**175**(1-2):213-228

[4] Gardner JP. The isopiestic method and the measurement of the osmotic and activity coefficients of aqueous electrolyte solutions. Journal of Solution Chemistry. 1988;**17**(11):911-942

[5] Gouveia AS, Tome LC, Marrucho IM. Density, viscosity, and refractive index of ionic liquid mixtures containing cyano and amino acid-based anions. Journal of Chemical & Engineering Data. 2016;**61**(1):83-93

[6] Stark A. Ionic liquid structure-induced effects on organic reactions. Ionic Liquids. 2010;**290**:41-81

[7] Magee JW. Measurement of Osmotic Coefficients by the Isopiestic Method [Thesis]. Berkeley: University of California; 1960

[8] Rogers RD, Seddon KR. Ionic liquids--solvents of the future? Science. 2003;**302**(5646):792-793

[9] Königsberger E, Königsberger LC, Hefter G, May PM. Zdanovskii's rule and isopiestic measurements applied to synthetic Bayer liquors. Journal of Solution Chemistry. 2007;**36**:1619-1634

Chapter 7

Molecular Dynamics Simulation and Investigation of Natural Gas Sweetening Using Pyridinium-Based Ionic Liquids

Fatemeh Moosavi

Abstract

This chapter investigates three ionic liquids (ILs), namely butyl pyridinium acetate ([BPy][AC]), butyl pyridinium benzoate ([BPy][BZ]), and butyl pyridinium propionate ([BPy][PR]), applied as potential absorbents for acid gases (hydrogen sulfide and carbon dioxide) in natural gas. The molecular dynamics (MD) simulation results indicate that the ILs have a relatively low dynamic and compact structure, with high viscosity in their pure state. Consistent with the findings of other researchers, the qualitative analysis of the simulation data for the mixture of an IL with acid and methane gases suggests that the dynamics of the IL enhances in the presence of these gases. The radial distribution functions reveal strong interactions and structural compatibility between the ILs and hydrogen sulfide molecules, indicating their suitability for hydrogen sulfide absorption. The amount of carbon dioxide gas absorbed by these ILs was calculated to be in the range of 0.08–0.11, while the absorption of hydrogen sulfide gas ranged from 0.12 to 0.18. [BPy][PR] IL exhibited the highest percentage of absorption for carbon dioxide (0.1083) and hydrogen sulfide (0.177). Furthermore, a comparison of the interactions between acidic gases and [BPy][PR] with the results of methyldiethanolamine (MDEA) clearly demonstrates the superior physical absorption of these gases by [BPy][PR].

Keywords: molecular dynamics simulation, gas sweetening, pyridinium based ionic liquids, adsorption, carbon dioxide

1. Introduction

Destruction of the environment has become one of the main obstacles to the further progress of society [1]. For example, large amounts of unwanted gases produced by various industries contribute to global warming, acid rain, and air pollution, posing serious threats to the human environment. Natural gas is a fossil fuel known for having fewer environmental effects compared to other fossil fuels. It primarily consists of methane, ethane, and trace amounts of propane and butane. Additionally, natural gas contains compounds like hydrogen sulfide and carbon dioxide, which must be removed due to their toxicity, corrosiveness, and ability to reduce the heat value of the fuel. Gas sweetening is a crucial process for purifying industrial gases, and various gas separation technologies, such as physical and chemical solvents, pressure swing adsorption, and membranes, are used in this field. However, the complexity of gas components and varying conditions often lead to high energy consumption, costs, and secondary pollution in most technologies. Two of the harmful gases that are detrimental to the environment and are classified as acid gases are carbon dioxide and hydrogen sulfide [2]. The most common method used in industries to absorb these gases is the use of aqueous amine solutions, such as monoethanolamine (MEA) [3–7], diethanolamine, methyldiethanolamine (MDEA) [8], and MDEA-piperazine mixtures [9]. However, issues such as corrosion, solvent degradation, high energy consumption, and pollution caused by volatility limit the effectiveness of these methods [10]. In recent years, a new type of solvents, known as green solvents or ionic liquids (ILs), have been introduced due to their exceptional properties for separating acid gases from natural gas [2, 11].

The process of removing unwanted compounds, such as hydrogen sulfide and carbon dioxide, from natural gas to reduce environmental and corrosion risks and increase the heat value of the fuel is known as natural gas sweetening. It is essential to separate these unwanted compounds to the fullest extent before using natural gas. The acceptable limits for hydrogen sulfide and carbon dioxide in natural gas are 4 ppm [12] and 2.5% [13], respectively.

In the oil and gas industry, amino solvents are used to separate acid gases from methane. These solvents have high absorption power due to chemical absorption but also drawbacks such as high vapor pressure, corrosion, and high energy consumption during solvent reclamation. For this reason, researchers are seeking solvents that can serve as alternatives to traditional solvents. One of the proposed alternatives are ILs, which have the ability to physically absorb significant amounts of acidic gases and can be considered as substitutes for amines. Compared to traditional solvents, ILs offer desirable properties such as high polarity; low volatility and inflammability; high thermal, chemical, and electrochemical stability; lower energy consumption; reduced absorbent and water use; the ability to dissolve many organic, inorganic, and organometallic compounds; modulated miscibility with various organic solvents; and adjustable structures [14].

Organic molten salts consisting of cations and anions with a melting point lower than 100°C are called ILs that are highly versatile as "design solvents." Their valuable physical and chemical properties include a low melting point (less than 100°C), very low vapor pressure, high thermal stability, high electrical conductivity, low surface tension, and adjustable viscosity. In separation processes, important parameters include the melting point, viscosity, vapor pressure, thermal stability, and chemical stability [15]. If the melting point of an IL is around room temperature, it is referred to as a room temperature ionic liquid (RTIL). These compounds typically consist of a large asymmetric organic cation paired with an organic or inorganic anion [16]. The significant interest in ILs arises from their ability to be designed as non-volatile substances. Furthermore, the wide range of available cations and anions, and their ability to be combined in various ways, offers great flexibility in processes. ILs can contain both polar and non-polar components, making them ideal for dissolving substances with different polarities [2]. Various cations and anions can be used to create ILs resulting in a diverse array of ILs tailored for specific applications or with enhanced physicochemical properties.

Common cations include imidazolium, pyridinium, pyrrolidinium, ammonium, phosphonium, and sulfonium [17, 18], and common anions include bis(trifluoromethylsulfonyl) imide, trifluoromethyl sulfate, dicyanamide, tetrafluoroborate, and hexafluorophosphate. Additionally, there are simple anions like chloride, bromide, iodide, nitrate, perchlorate, formate, and acetate, which may have less stability or may not be liquid at room temperature.

Some physical properties of ILs, such as viscosity, boiling point, and solubility in water, can be adjusted based on the ion pair selection or through the introduction of special functional groups into the structure of the cation or anion. In general, the most important advantages of ILs are as follows:

1. These compounds have low vapor pressure, making them non-volatile and environmentally friendly compared to common organic solvents.

2. ILs are more conductive than organic compounds and can easily dissolve various enzymes.

3. These green solvents have high thermal stability and do not decompose at high temperatures, making product separation easy.

4. ILs have low flammability, making them safe for chemical reactions.

5. ILs can dissolve various compounds due to their structural diversity, and their thermodynamic properties can be improved by changing the structure.

6. These compounds have high density and polarity, remaining liquid over a wide range of temperatures.

ILs can be categorized into two important groups: conventional ILs and task-specific ILs [3]. The key difference is that task-specific ILs have chemical absorption capabilities, able to absorb acid gases with low concentrations, while conventional ILs primarily perform physical absorption. When the concentration of acid gas in the input feed is low, conventional ILs do not have a high absorption capacity [19]. Research has shown that functional ILs are capable of both physical and chemical absorption. In some functional ILs, certain atoms, groups, or structures of cations exist as functional groups to absorb acid gases. For example, 1-(2-diethylaminoethyl)-3-methylimidazolium hexafluorophosphate and 1-(2-diethylaminoethyl)-1-methylpyrrolidinium hexafluorophosphate are functional ILs used for the absorption of sulfur dioxide due to the amine group in the cation structure [20].

Although imidazolium-based ILs have disadvantages such as high manufacturing cost, viscosity, and toxicity [21], the low cost [3], low toxicity, high thermal stability, and biodegradability of ILs based on pyridinium cation make them favorable.

The ability to create a wide variety of ILs has led to the use of molecular dynamics (MD) simulation and computational chemistry methods to calculate the electronic, thermodynamic, and phase equilibrium properties of ILs [22–35].

Additionally, the solubility of a mixture of CO_2 and H_2S in aqueous solutions of amines and ILs, such as diisopropanolamine (DIPA) blended with [bmim][acetate], has been measured [2]. The gas solubility is influenced by the concentration of DIPA, with the addition of IL leading to increased solubility. Pyridinium-based ILs have excellent properties including low cost, viscosity, and toxicity; high thermal stability and biodegradability; and relatively high CO_2 solubility according to Frias et al. [11].

Sazanova et al. [1] focused on amphiphilic bis(2-ethylhexyl) sulfosuccinate anion and imidazolium, pyridinium, and pyrrolidinium cations to determine H_2S and CO_2 solubility in all prepared ILs. They demonstrated that van der Waals (vdW) interactions, in addition to electrostatic and hydrogen bonding, play a significant role in acid gas removal by these ILs, leading to the conclusion of a physical absorption mechanism. The protic nature of the cation results in a notable reduction in gas solubility, and the sorption capacity is influenced by the aprotic nature of the cation.

Wang et al. [36, 37] synthesized a series of pyridinium-based ILs and investigated their properties including density, viscosity, thermal decomposition, and gas separation performance for H_2S, CO_2, and SO_2. The results showed that the solubility of gases in ILs increases significantly with pressure and decreases with temperature. Additionally, as the length of the alkyl chain attached to the cation increases, the solubility rises due to the intensity in vdW force between the ions and the acid gas. Among the studied ILs with the same cations, the [C_4Py][SCN] exhibited the highest selectivity about 1.5–4 times larger than imidazolium-based ILs. These ILs showed a greater molar fraction of hydrogen sulfide gas absorbed compared to CO_2. Furthermore, pyridinium-based ILs demonstrated stable absorption performance after five consecutive cycles, indicating their potential as effective absorbents for gas separation applications.

The direct emission of ammonia gas by the chemical industry severely pollutes the environment. Traditional methods for removing this gas include washing with water or acid. Nevertheless, absorption by ILs offers advantages such as no solvent wastage and easy recovery of NH_3 [38–41]. Yokozeki et al. [42, 43] introduced eight types of ILs for absorbing ammonia gas at different temperature and pressure ranges, suggesting a complex molecular interaction between ammonia gas and ILs. Yuan et al. [40] stated that NH_3 gas interacts chemically and physically with dual-functionalized pyridinium-based ILs containing acidic protons and hydroxyl groups. Further studies on six-membered N-heterocyclic cations in protic ILs (PILs) confirmed thermal stability during NH_3 absorption with high capacities and selectivity [41]. Pyridinium- and piperidinium-based PILs were identified as acceptable candidates for NH_3 absorption due to their better biodegradability as well as lower cost compared to imidazolium-based ILs, with the acid-base interaction and hydrogen bonding [44] playing a crucial role in NH_3 interaction with IL.

Sulfur dioxide is another acid gas present in exhaust gases from power plants. Various ILs have been designed and synthesized as efficient absorbents for the absorption and separation of SO_2, offering opportunities for new separation processes with easy recovery, high capacity, and low enthalpy of absorption [20]. Wang et al. [45] have shown that small-sized ILs with moderate basicity of anions are more preferential for SO_2 absorption.

Zeng et al. [36, 46, 47] investigated the effect of cation alkyl chain length and anion type on the solubility of sulfur gas in ILs, finding that anions play a significant role in the absorption of sulfur dioxide gas. Functionalized ILs with ether and nitrile groups exhibited high absorption capacity due to strong physical interactions with sulfur dioxide gas compared to conventional ILs [47, 48].

Shang et al. [10] reported that the chemical and physical absorption of SO_2 gas affects the viscosity of pyridinium-based ILs, with chemically absorbed gas leading to an increase in viscosity. While most studies in the literature have focused on imidazolium and ammonium cations, investigations on other forms of cations are needed. Pyridinium-based ILs have shown promise for acidic gas capture, with ongoing research focusing on natural gas sweetening and selectivity of these ILs paired with carboxylate anions. The study aims to investigate the structures and properties of ILs and their efficiency in the natural gas sweetening process, highlighting their high absorption capacity and thermal stability, low vapor pressure, and adjustable structures that make them ideal for absorbing and separating acid gases.

2. Simulation details

In this study, MD simulations were conducted to investigate the removal of acid gases, including hydrogen sulfide and carbon dioxide from natural gas using ILs with butylpyridinium cation ([BPy^+]) and acetate ([AC^-]), benzoate ([BZ^-]), and propionate ([PR^-]) anions at 325 K and 1 atm. The temperature and pressure of the systems were maintained constant using the Nose-Hoover thermostat and barostat, respectively. The simulations were carried out using DL_POLY software version 2.17 [49], and the results were analyzed using VMD software version 1.9.1 [50]. MD simulations were performed for various systems, including ILs in a liquid and pure state, IL-carbon dioxide, IL-hydrogen sulfide, IL-CO_2-H_2S, and IL-CO_2-H_2S-CH_4 mixtures. The simulation process consisted of three stages: preparation of input files, conducting the simulation, and extracting and analyzing the results.

To initiate the simulation, the initial configuration of the system was selected, and an initial cubic simulation box with an edge of 150 Å containing 125 ion pairs of IL and a specific number of gas molecules (90 molecules of methane, 4 molecules of H_2S, and 6 molecules of carbon dioxide) was prepared. Information about the force fields for the cation $[BPy]^+$ [51]; anions $[AC]^-$ [52], $[BZ]^-$ [53], and $[PR]^-$ [54]; and the target gases CO_2 [53], H_2S [55], and CH_4 [39, 56] was obtained from reliable scientific sources. Simulations were conducted with a time step of 1 fs. The system first equilibrated in NPT ensemble for 1 ns, followed by production runs in the NVT ensemble for 1 ns.

After preparing the input files, simulations were conducted with the NPT statistical ensemble, where the system volume was equilibrated and then fluctuated around a constant average value. Subsequently, simulations were continued in the NVT ensemble to collect results. Once the simulations were completed, the results were analyzed to calculate the structural and dynamic properties of the system. **Figure 1** shows the structure of each anion and cation studied in the present study.

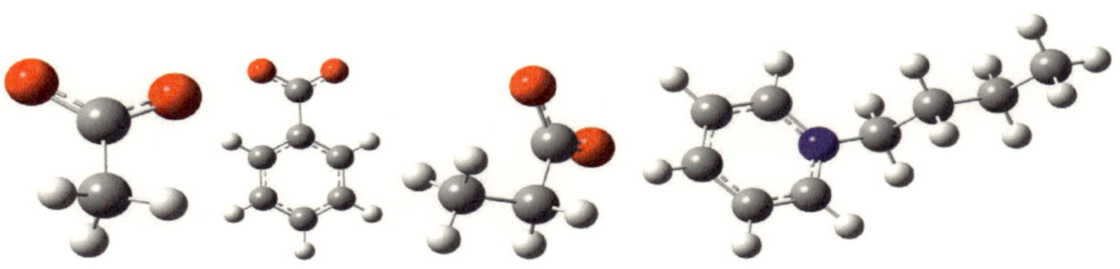

Figure 1.
The structure of studied anions (from the left $[AC]^-$ (first), $[BZ]^-$ (second), and $[PR]^-$ (third)) and cation $[BPy]^+$ (right).

3. Results and discussion

3.1 Pure ILs

After equilibration, the ratio of fluctuation to the root mean square (rms) of energy and volume quantities for pure ILs was computed. If the fluctuation in rms is less than 0.01, the system is stable, and the simulation time is sufficient. For [BPy][AC], [BPy][BZ], and [BPy][PR] ILs, the ratios of fluctuation to rms in energy values are 0.183, 0.178, and 0.247 J, respectively. The volume fluctuation ratios are

0.0045, 0.00153, and 0.00321 Å^3, respectively. These results suggest that 1 ns of MD simulation is adequate.

Accurate thermophysical properties, like intermolecular energy, are essential for fluid modeling. Some properties are challenging to calculate in labs but can be easily evaluated through MD simulations [57]. **Table 1** lists of some of these properties at 325 K and 1 atm.

Comparing intermolecular energy in three simulated pure ILs (**Table 1**) reveals that [BPy][BZ] has the strongest U_{int}. Electrostatic energy strongly influences the system's energy, explaining ILs' unique properties like low volatility and vapor pressure due to high electrostatic energy. [BPy][AC] has the lowest vdW energy, attributed to the acetate anion's small size. Larger particles have considerable vdW energy, explaining why [BZ]$^-$ has the highest U_{vdW}.

Mean square displacement (MSD) increases over time provide insight into species dynamics. MSD changes with time for anions and cations in each pure system at 325 K were calculated. The graphs in **Figure 2** show almost linear behavior, with self-diffusion coefficients (D) calculated from the linear part's slope. D values range from 100 to 400 picoseconds and are listed in **Table 2**.

Self-diffusion coefficients for anions and cations are around 10^{-11} m^2/s, consistent with experimental values [58–60]. Anions have faster dynamics and D values than cations due to their smaller sizes and mass, reflecting faster mobility [58]. Density comparisons validate the simulation, with [BPy][AC] showing acceptable agreement [57]. Viscosity was calculated using Stokes-Einstein Eq. (1) [61]:

$$D_{IL} = D_{H_2O} \frac{\eta_{H_2O}}{\eta_{IL}} \tag{1}$$

The density and viscosity values are given in **Table 2**.

IL	Intermolecular energy (U_{int})	Electrostatic energy (U_{ele})	vdW energy (U_{vdW})
[BPy][AC]	−68,230	−62,516	−5714
[BPy][BZ]	−75,677	−67,788	−7889
[BPy][PR]	−73,810	−68,019	−5791

Table 1.
Comparison of intermolecular, electrostatic, and van der Waals (vdW) energies in terms of kJ/mol in each of the pure [BPy][AC], [BPy][BZ], and [BPy][PR] ILs.

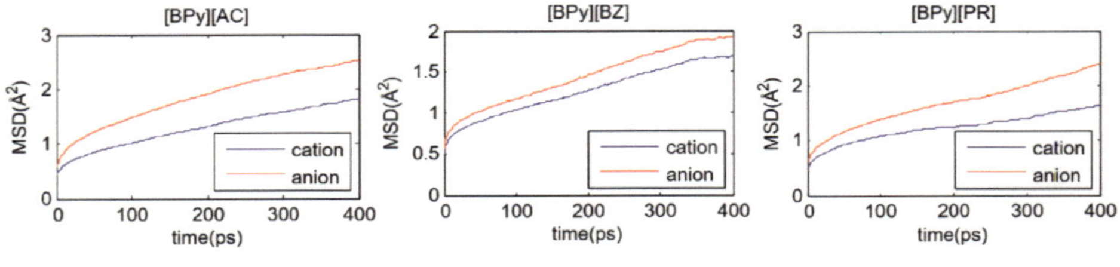

Figure 2.
MSD variation with simulation time for studied pure ILs.

IL	ρ_{Sim} (g/cm^3)	D_+ (m^2/s)	D_- (m^2/s)	η_{IL} (cP)
[BPy][AC]	0.996	1.100×10^{-11}	1.565×10^{-11}	136
[BPy][BZ]	0.840	1.013×10^{-11}	1.154×10^{-11}	167
[BPy][PR]	0.968	1.008×10^{-11}	1.351×10^{-11}	153

Table 2.
Density, D of cations (D_+), D of anions (D_-), and viscosity of pure ILs at 325 K.

Radial distribution functions (RDFs) provide detailed insight into the IL structure. RDFs of anions around the cation's HA1 atom show strong hydrogen bonding (HB) (**Figure 3**). Benzoate anion has the most interaction with $[BPy]^+$ cation, forming a compact structure with strong intermolecular forces. Center of mass (COM) RDFs for cation-cation, anion-cation, and anion-anion pairs (**Figure 4**) show different interactions. Anion-cation interactions are strong, with approximately six anions per cation in each IL, consistent with prior research [57].

3.2 Mixed systems

After studying the dynamic and structural properties of pure ionic liquids [BPy][AC], [BPy][BZ], and [BPy][PR], the systems of these ILs with CO_2 and H_2S were separately examined. The simulation systems reviewed in this section include a mixture of IL with CO_2 gas, a mixture of IL with H_2S gas, a mixture of IL with a

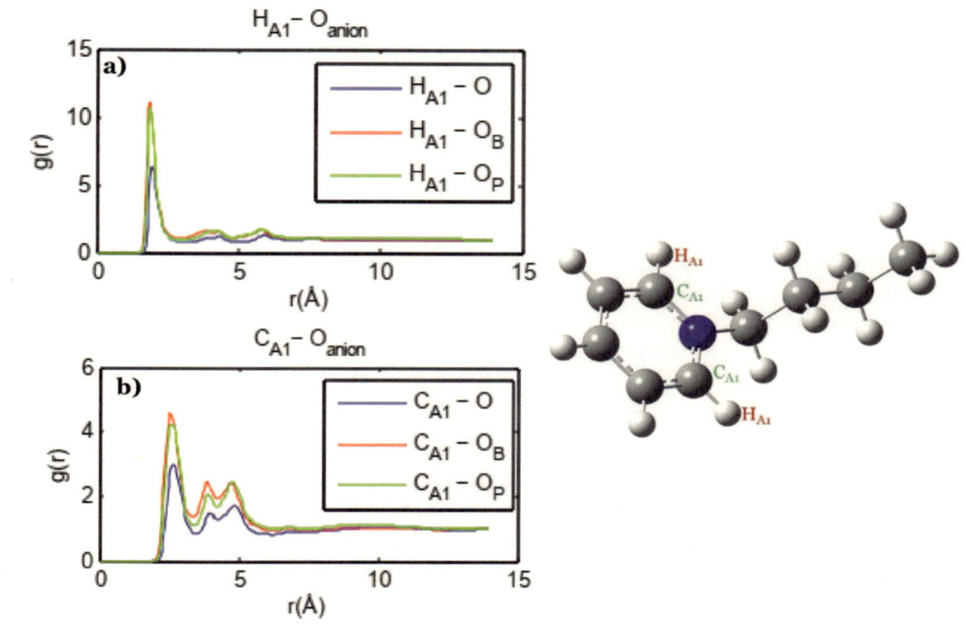

Figure 3.
RDF of anion oxygen atom around cation (a) H_{A1} atom (shown in legend), (b) C_{A1} atom in each pure system of ionic liquids [BPy][AC] (blue), [BPy][BZ] (red), and [BPy][PR] (green).

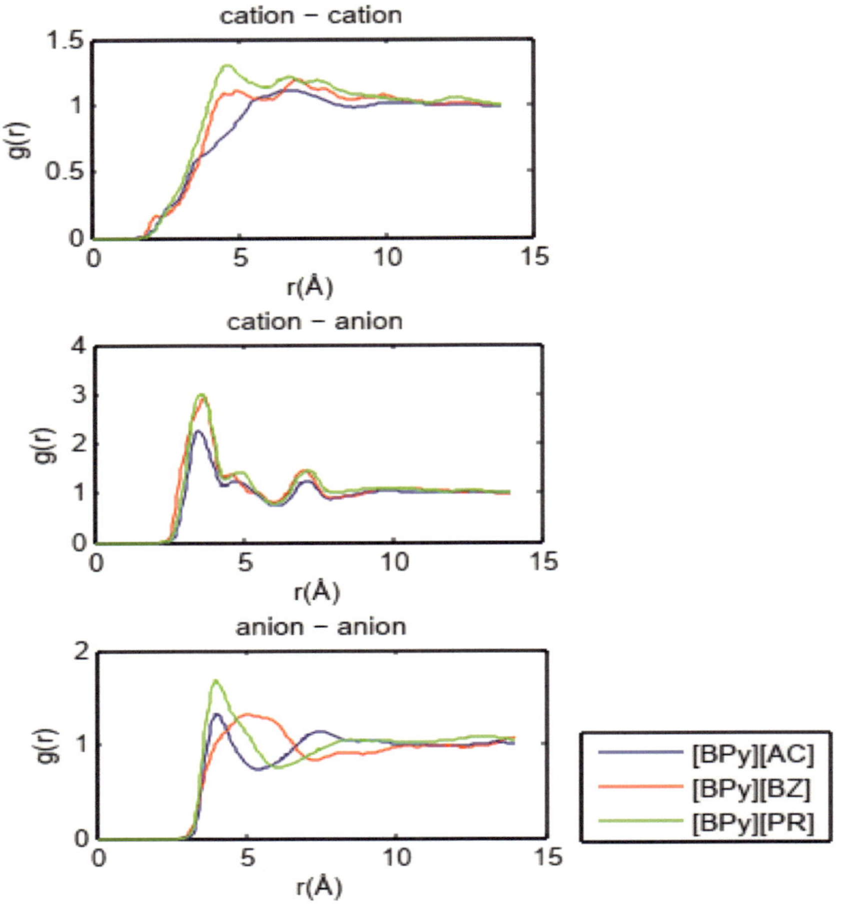

Figure 4.
Diagram of RDFs at 325 K for cation-cation, cation-anion, and anion-anion of pure studied ILs.

combination of CO_2 and H_2S gases, and finally a mixture of IL with a combination of CO_2, H_2S, and CH_4 gases. The dynamic and structural properties of these systems are presented and analyzed.

Figure 5 compares the MSD for $[BPy]^+$ cation and $[AC]^-$, $[BZ]^-$, and $[PR]^-$ anions in pure and mixed systems with IL + CO_2.

As shown in **Figure 5**, the dynamics of the cation and anions in the mixed systems increase significantly compared to the pure state, indicating a strong interaction between ions and carbon dioxide gas. This finding aligns with previous research by Morganti et al. [62] who studied the solubility of SO_2 and CO_2 gases in ammonium-based ILs and found that gas solubility increased the *D* of IL species.

Figure 6 illustrates the comparison of the MSD of cations and anions in all three mixed systems. The dynamics of anions and cations in the [BPy][AC] system with CO_2 gas is higher than in the other two systems, likely due to the small size of acetate anion and the cluster movement of ILs.

Figure 7 displays the MSD of CO_2 in mixtures with CO_2 gas and [BPy][AC], [BPy][BZ], and [BPy][PR] ILs.

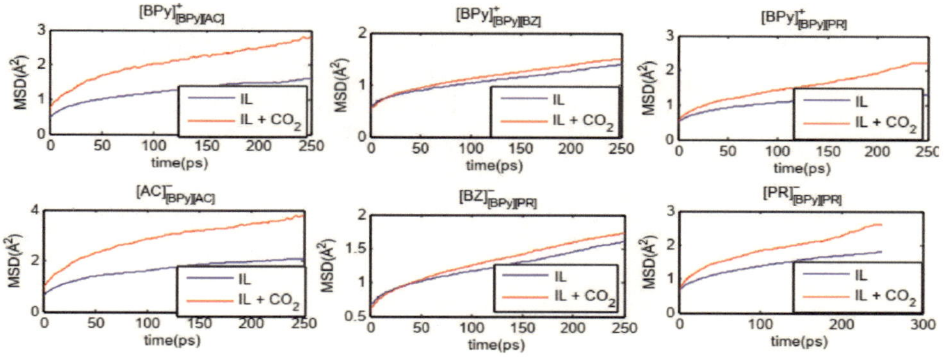

Figure 5.
Comparison of the MSD for cation and anions between the pure IL and mixed with CO_2 at 325 K.

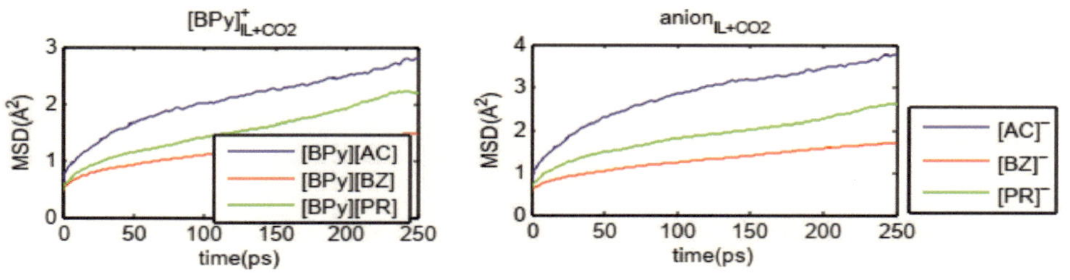

Figure 6.
Comparison between MSD of cation and anion in each mixed system with CO_2 and pure ILs at 325 K.

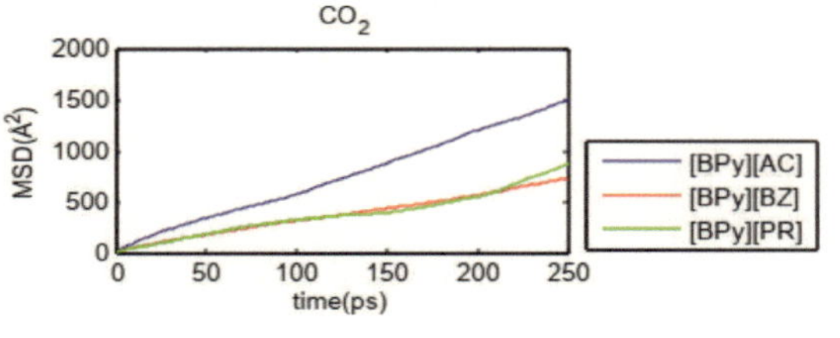

Figure 7.
Comparison between MSD of CO_2 in the target ILs [BPy][AC], [BPy][BZ], and [BPy][PR].

The dynamics of CO_2 molecules in the [BPy][AC] + CO_2 system is significantly different from the other systems, possibly due to the low absorption rate of CO_2 molecules in [BPy][AC] IL, resulting in a separate gas phase within the simulation box; see **Figure 8** for more details.

The diffusion coefficients of each species were calculated using Einstein's relation [63], with the values reported in **Table 3**.

According to **Table 3**, the self-diffusion coefficient of CO_2 gas dissolved in IL is larger than the corresponding values of ions in good agreement with Morganti and the co-workers' research [62]. The D_{CO2} in [BPy][BZ] IL is the lowest due to the large size of benzoate anion and strong intermolecular, electrostatic, and vdW forces in this system (see **Table 1** for more details), leading to a compact structure with high

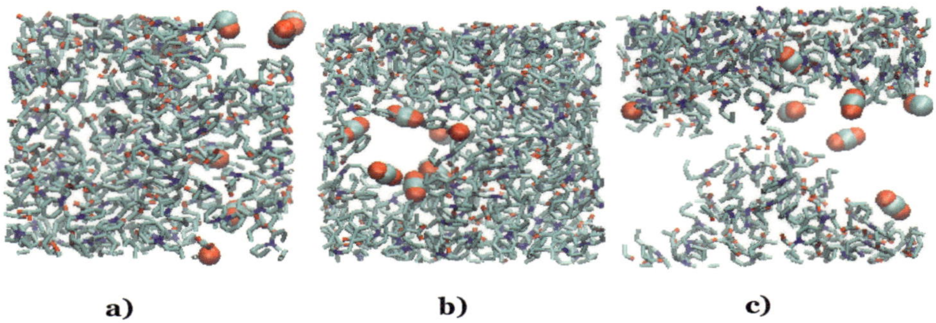

Figure 8.
Graphical representation of the CO_2 absorption in (a) [BPy][AC], (b) [BPy][BZ], and (c) [BPy][PR]. Red, blue, and dark blue colors display oxygen, carbon, and nitrogen atoms, respectively.

IL + CO_2	D_+ (m^2/s)	D_- (m^2/s)	D_{CO2} (m^2/s)
[BPy][AC]	2.030×10^{-11}	2.778×10^{-11}	1.020×10^{-8}
[BPy][BZ]	1.095×10^{-11}	1.136×10^{-11}	0.475×10^{-8}
[BPy][PR]	1.721×10^{-11}	2.024×10^{-11}	0.485×10^{-8}

Table 3.
Cation, anion, and CO_2 gas diffusion coefficient in the mixture containing CO_2 gas and each IL at 325 K.

viscosity. The role of anion in gas absorption is examined through the RDF of the COM of each species. **Figure 9** compares the RDF of cation-CO_2 and anion-CO_2 to explore the strength of interaction with the absorbed gas.

The RDF analysis shows that anion interacts more with the absorbed gas than the cation, with $[PR]^-$ anion displaying the strongest interaction with CO_2 gas among the three ILs. From the other side of view, the terminal atoms related to the anion chain also show a strong interaction with CO_2 molecules, that is, stronger than that of cations. This observation is in agreement with the literature [64–66].

According to the observations of Ren et al. [67], if the acidic pKa of the anion corresponding to the acids that make up the ILs is greater than 6.36, the dissolution of CO_2 gas in this type of ILs takes place in the form of chemical absorption. Otherwise, CO_2 absorption is done physically. In fact, the higher the pKa, the greater the alkalinity of ILs, the greater the adsorption capacity while the selectivity decreases. The nature of the anion mother is acidic. The pK_a value for acetic acid is equal to 4.75, in the case of benzoic acid $pK_a = 4.202$, and the value for propionic acid is 4.88 [68]. As can be seen, the pK_a of these acids is less than 6.36. Therefore, it can be expected that the interactions of these ILs with carbon dioxide gas are weak, which is indicated by the presence of broad peaks with low intensity in **Figure 9**. It can also be expected that the $[PR]^-$ anion has a strongest interaction with CO_2 gas in comparison with $[AC]^-$ and $[BZ]^-$ anions. The existence of a peak with high intensity compared to the other two anions confirms this result.

Another expectation is that the intensity of interaction between CO_2 and $[BZ]^-$ anion is the lowest since it has the least alkaline property. However, it is observed that $[BZ]^-$ anion-CO_2 intensity is almost similar to $[AC]^-$ anion-CO_2 interaction. This is due to the large size of $[BZ]^-$ anion. In fact, physical absorption is influenced by vdW and electrostatic forces. The larger the size of the ions, the stronger the vdW force between the species [57].

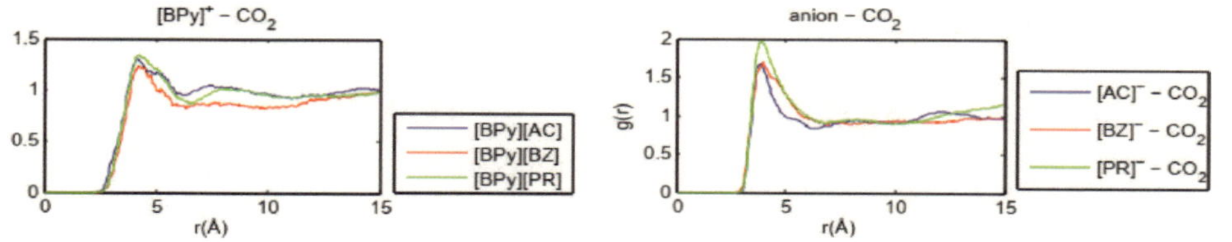

Figure 9.
The RDF between CO_2 and IL ion pairs at 325 K in [BPy][AC] + CO_2, [BPy][BZ] + CO_2, and [BPy][PR] + CO_2 systems.

IL + CO_2	N_+	N_-
[BPy][AC]	0.0404	0.0415
[BPy][BZ]	0.0423	0.0552
[BPy][PR]	0.0432	0.0645

Table 4.
Coordination number of CO_2 around the cation (N_+) and each anion (N_-) in the mixed system including CO_2 gas with ILs [BPy][AC], [BPy][BZ], and [BPy][PR].

The coordination number of CO_2 gas around anion and cation is also reported in **Table 4** showing CO_2 molecules are most distributed around [BPy][PR] IL, indicating high absorption due to its alkalinity.

Figure 10 compares the MSD of pure and mixed systems with H_2S gas and ILs [BPy][AC], [BPy][BZ], and [BPy][PR].

The MSD analysis of H_2S gas absorption in ILs presents increasing dynamics of ions compared to pure IL. [BPy][AC] IL shows the highest dynamics for the small size of the acetate anion. The coordination number of H_2S molecules around each IL is reported in **Table 5**. The table reveals that hydrogen sulfide is most absorbed by [BPy][PR] IL, attributed to its higher alkalinity. The RDF analysis of H_2S absorption in ILs (**Figure 11**) shows stronger interaction with the cation than the anion, with $[PR]^-$ anion exhibiting similar intensity to the cation due to its high alkalinity.

Radial distribution function (RDF) investigation shows that H_2S gas accumulates around the cation alkyl chain and anion, with strong interactions observed at the terminal atoms of this chain. The results of other researchers confirm this finding [69]. The polarity of hydrogen sulfide is lower than that of water, resulting in higher solubility in organic solvents compared to aqueous solvents [69]. This can be attributed to the less polar nature of alkyl chain if compared to water, leading to better absorption in organic solvents. Therefore, hydrogen sulfide molecules are absorbed in the organic compartment of cation and anion.

As depicted in **Figure 11**, the interaction between the cation and H_2S is stronger than the anion and hydrogen sulfide. This is due to the larger size of the cation relative to the anion. According to the literature [69, 70], the large molecular volume is responsible for gas solubility. Therefore, as the $[BPy]^+$ cation is larger than the $[AC]^-$ and $[BZ]^-$ anions, H_2S molecules are more concentrated around the cation, resulting in stronger interactions with the cation than with the anion. Analysis of the RDF related to the mixed system containing hydrogen sulfide and [BPy][PR] IL reveals that despite the cation's larger size compared to the $[PR]^-$ anion, the intensity of the molecular interaction of hydrogen sulfide with $[PR]^-$ anion is nearly equivalent to its interaction with the cation. This could be attributed to the high alkalinity of the $[PR]^-$ anion.

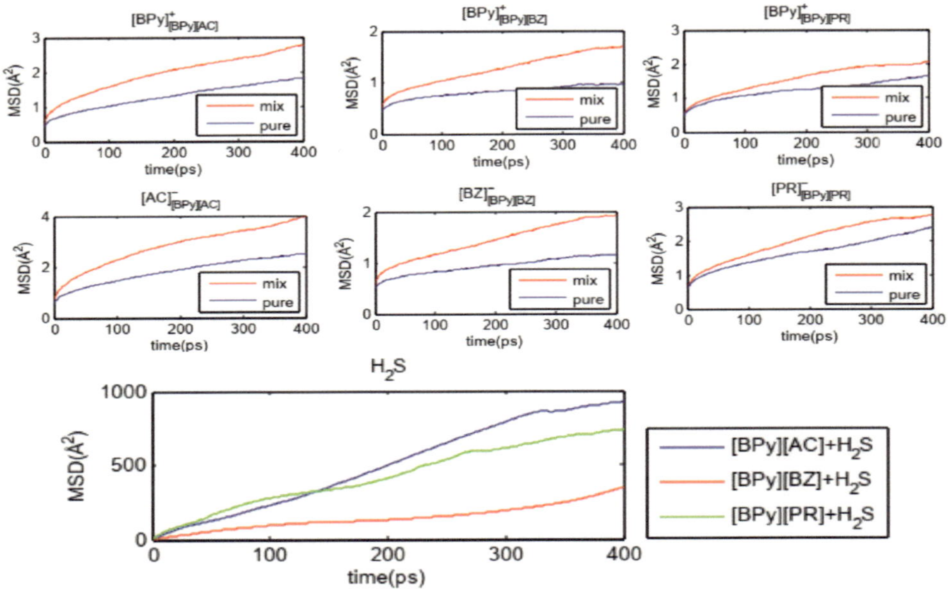

Figure 10.
Comparison of the cation, anion, and H_2S MSD in pure and mixed states with H_2S gas in [BPy][AC], [BPy][BZ], and [BPy][PR] ILs.

IL + H_2S	D_+ (m^2/s)	D_- (m^2/s)	D_{H2S} (m^2/s)	N_+	N_-
[BPy][AC]	1.603×10^{-11}	2.320×10^{-11}	0.413×10^{-8}	0.055	0.062
[BPy][BZ]	1.187×10^{-11}	1.243×10^{-11}	0.120×10^{-8}	0.052	0.083
[BPy][PR]	1.270×10^{-11}	1.665×10^{-11}	0.353×10^{-8}	0.056	0.121

Table 5.
Cation, anion, and H_2S gas diffusion coefficient values and coordination number of H_2S around $[BPy]^+$, $[AC]^-$, $[BZ]^-$, and $[PR]^-$ in the mixture of IL and H_2S at 325 K.

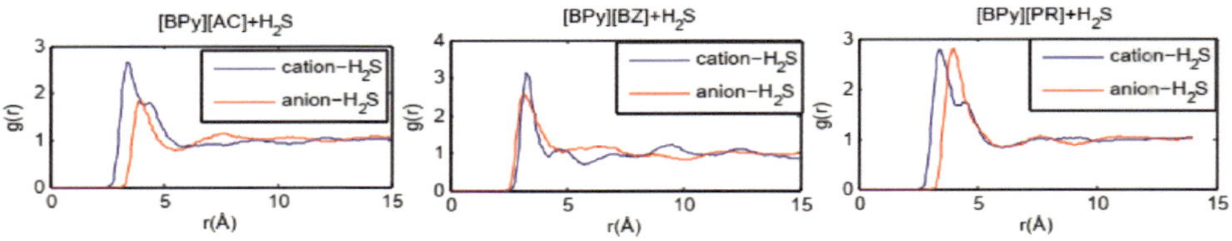

Figure 11.
The RDF of H_2S around cation and anion of [BPy][AC], [BPy][BZ], and [BPy][PR].

Careful examination of the graphical representation of each IL + H_2S system (**Figure 12**) reveals that hydrogen sulfide molecules are more readily absorbed on the surface of the IL. This results in a reduction of ion-ion interaction due to the interaction of ions with the gas molecules, ultimately leading to increased mobility and movement of ions within the mass. The fast diffusion of ions in mixed systems, as opposed to pure ones, is attributed to the strong interaction between ions and

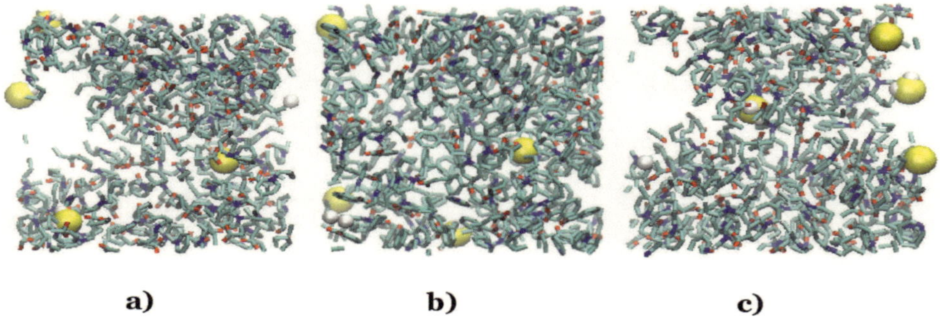

Figure 12.
Graphical representation of the H_2S absorption in (a) [Bpy][AC], (b) [Bpy][BZ], and (c) [Bpy][PR]. The colors red, light blue, dark blue, and yellow represent oxygen, carbon, nitrogen, and sulfur atoms, respectively.

hydrogen sulfide. Specifically, H_2S penetrates more into the [BPy][BZ] IL compared to the other two systems, resulting in these molecules entering the bulk of the IL, decreasing mobility more than the other two systems. In summary, the larger the size of the ion, the stronger the interaction with the absorbed gas, highlighting the importance of anion properties in gas absorption in ILs.

3.3 Mixed systems and selectivity (mixed gases)

The investigation of the selectivity and structural properties of the mixed system containing H_2S and CO_2 gases with each IL [Bpy][AC], [Bpy][BZ], and [Bpy][PR] was conducted. Selective separation of H_2S and CO_2 gases is crucial in natural gas sweetening. Therefore, it is essential to use solvents with high selectivity for H_2S gas over CO_2 in natural gas sweetening. As mentioned above, the larger the size of the ions and the greater the alkalinity of ILs, the more acidic gases (H_2S and CO_2) are absorbed, but the selectivity of H_2S decreases. Therefore, ILs with small ions and low alkalinity may exhibit high selectivity.

In **Figure 13**, the RDF diagram of H_2S and CO_2 molecules around cations and anions in each of the three-component mixed systems of H_2S + CO_2 + [Bpy][AC], H_2S + CO_2 + [Bpy][BZ], and H_2S + CO_2 + [Bpy][PR] is depicted. The intensity of H_2S...cation interaction is greater than the intensity of H_2S...anion interaction. The absorption of hydrogen sulfide gas is more influenced by the size of the ions, with larger ions absorbing more hydrogen sulfide molecules. Increasing the length of the cation chain or using larger anions can enhance the absorption of H_2S gas. However, high alkalinity is also an important factor in the absorption of these molecules [69], as seen in the strong interaction of propionate anion with hydrogen sulfide (**Figure 11**).

Figure 13 also indicates that the cation-hydrogen sulfide interaction is stronger than the cation-carbon dioxide interaction. Enlarging the length of the alkyl chain improves the absorption of both gases. The interaction of anion-H_2S is stronger than that of anion-CO_2, with increased alkalinity leading to decreased selectivity [62]. The benzoate anion shows the highest selectivity due to its lower alkalinity compared to the other anions.

Comparing RDFs related to hydrogen sulfide and carbon dioxide around cations and anions in each of the mixed systems with ILs [BPy][AC], [BPy][BZ], and [BPy][PR] reveals consistency in each system. The intensity of the interaction between benzoate anion and hydrogen sulfide is stronger than that of acetate anion, attributed to

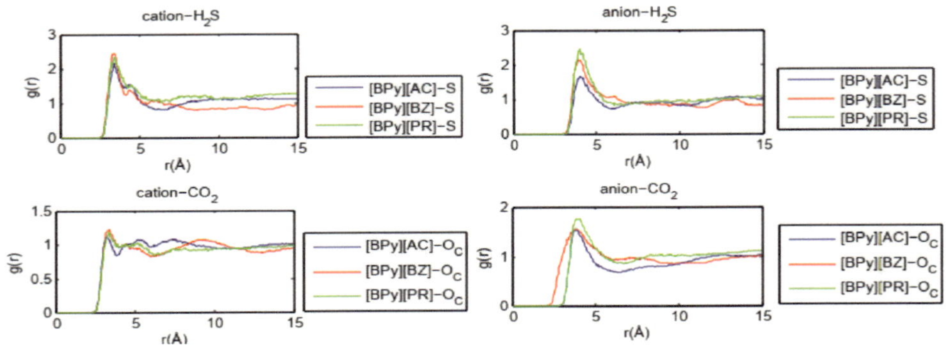

Figure 13.
RDF related to the S atom of H_2S and the O atom of CO_2 (O_C) around $[Bpy]^+$, $[AC]^-$, $[BZ]^-$, and $[PR]^-$ in each of the mixed systems containing both gases and IL.

the larger size of benzoate anion. The alkalinity of the anion influences the absorption of CO_2, with higher alkalinity leading to greater interaction and absorption.

Figure 14 presents the selectivity of H_2S gas over CO_2 and methane in each of the mixed systems.

The interactions of cation and anion with H_2S molecules are stronger than with other gases, CO_2 and CH_4, reflecting the higher solubility of H_2S in ILs. The increase in anion alkalinity enhances interaction with acid gases, that is, the strongest in the case of [BPy][PR]. However, increasing alkalinity can reduce selectivity [69].

Figure 15 depicts the MD process in a comprehensive global view. The first step involves simulating the pure IL. Next, CO_2 acidic gas is introduced into the IL, and the simulation begins. It is evident from the demonstration that CO_2 is fully dissolved in the [BPy][PR] IL. In the third step, a mixture of CO_2 and H_2S acidic gases is introduced into the solvent for MD simulation to determine if the gases are absorbed. The final step (4) includes not only the IL but also a mixture of CO_2, H_2S, and CH_4 gases. The illustration indicates that the CO_2 gas has successfully dissolved completely.

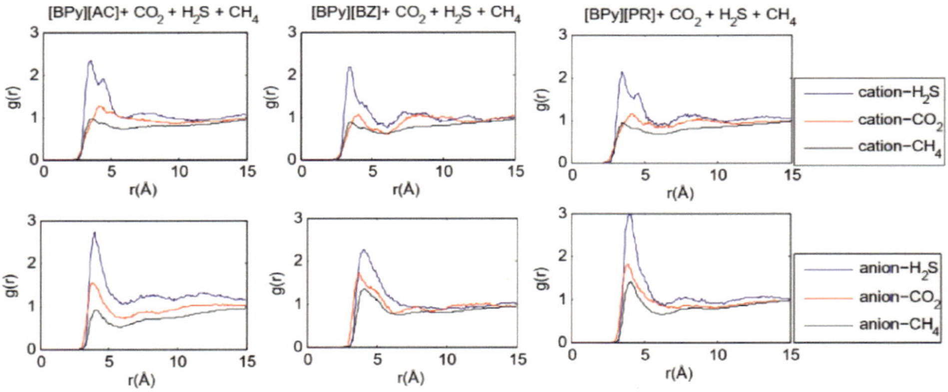

Figure 14.
The qualitative comparison of the selectivity of H_2S gas over CO_2 and CH_4 in each of the mixed systems including $[BPy][AC] + H_2S + CO_2 + CH_4$, $[BPy][BZ] + H_2S + CO_2 + CH_4$, and $[BPy][PR] + H_2S + CO_2 + CH_4$.

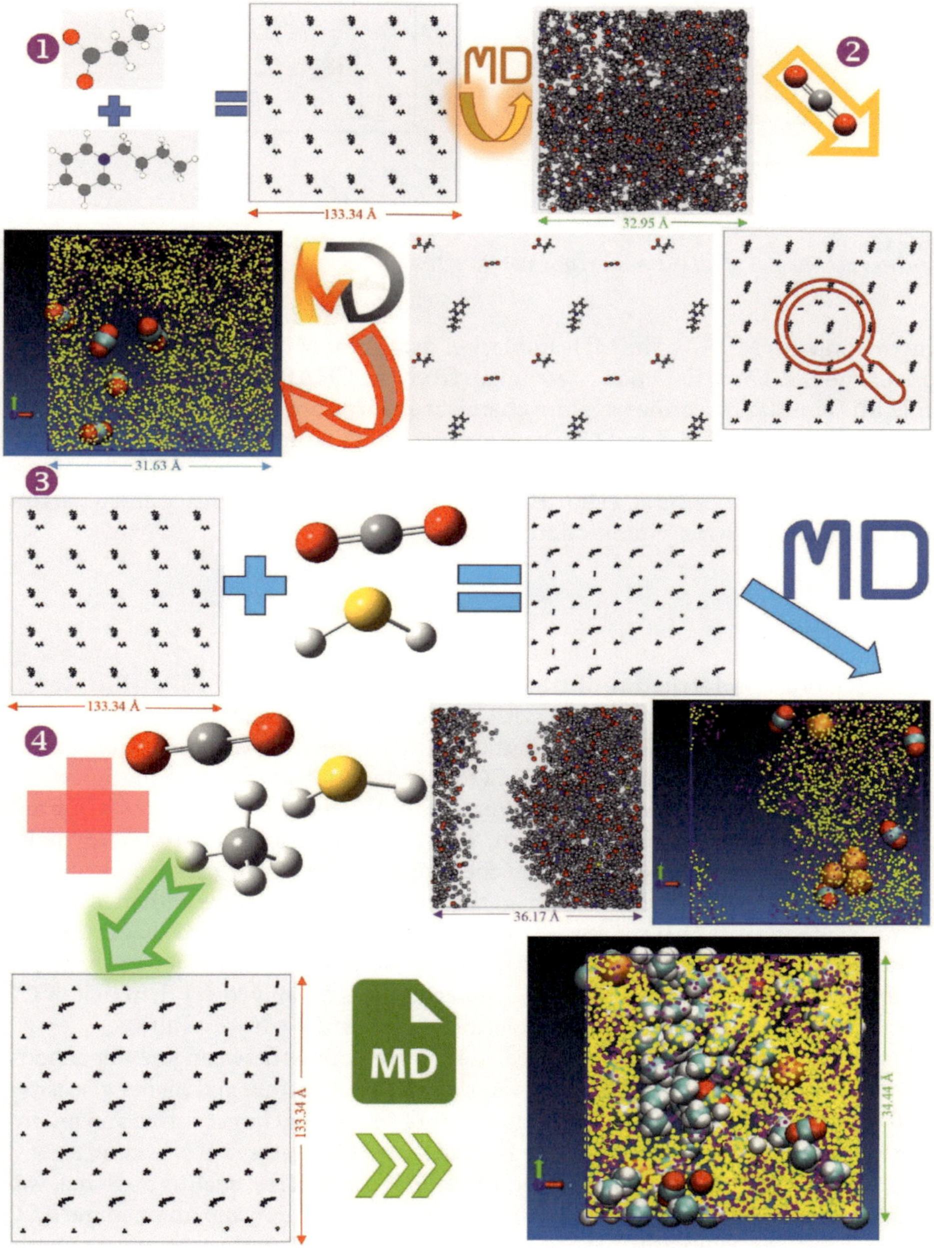

Figure 15.
The MD process for gas absorption by [BPy][PR] as a typical sample of the present study.

3.4 Comparison of the solubility of acid gases in [BPy][PR] and MDEA

In **Figure 16**, the solubility of H_2S and CO_2 gases in [BPy][PR] IL and MDEA amine solvent is compared. The results show that [BPy][PR] IL is the most effective solvent among the studied ILs for acid gas absorption, leading to its selection for

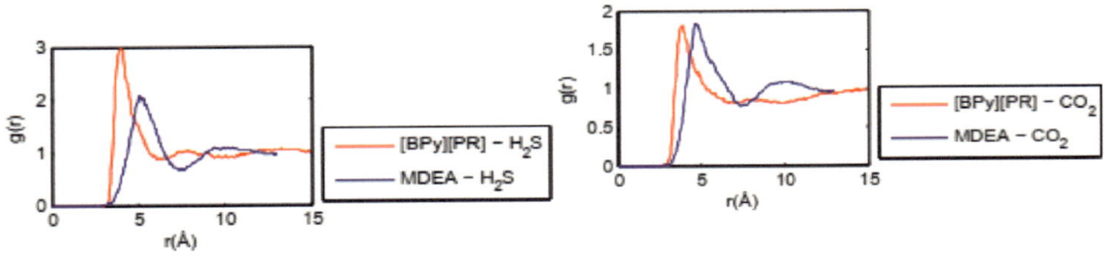

Figure 16.
Comparing the solubility of acid gases in MDEA and [BPy][PR].

comparison with MDEA. The MD simulation details for MDEA are consistent with those of the ILs. Force field data were taken from the CHARAM force field [71]. While MD simulation cannot explore chemical absorption, it does determine the physical absorption of H_2S and CO_2 gases, as depicted in **Figure 16**. The intensity of H_2S interaction with the IL is notably higher than with amine solvent. The peak in the RDF of IL with H_2S occurs at a shorter distance, with greater intensity and a narrower width, indicating a strong interaction with H_2S.

4. Conclusion

The dynamics and structure of ILs [BPy][AC], [BPy][BZ], and [BPy][PR] were studied in their pure state and mixed with H_2S, CO_2, and CH_4 gases. To analyze the dynamics of the systems, the MSD was utilized, while the RDF was used to examine the structure. The MSD quantity for the COM of the cation and each of the anions at 325 K was calculated to investigate the dynamics of the species. In all systems, the anion exhibited faster dynamics than the cation, which can be attributed to the lower molecular mass of the cation. The self-diffusion coefficient values for the cation and anion of pure ILs were calculated in each system at the specified temperature with values around 10^{-11} m^2/s for ion pairs of ILs. Results from the MSD analysis indicated that the presence of acid gases significantly increased the dynamics of ions due to interactions between ions and gas molecules. The MSD of H_2S molecules was notably lower than that of CO_2 molecules, likely due to the strong interaction of ions with hydrogen sulfide.

Analysis of the RDF for pure ILs revealed that interactions between ions depended on their size, with larger ions leading to a more compact structure and higher viscosity. When examining RDFs in mixed systems, it was observed that larger cation and anion sizes resulted in increased absorption of H_2S, CO_2, and CH_4 gases in ILs, leading to decreased selectivity. The solubility of poorly soluble gases in ILs, such as methane, was found to be more dependent on the volume and size of the cation and anion in the IL. On the other hand, the solubility of H_2S and CO_2 gases depended on electrostatic interactions in addition to the volume and size of ions. Higher alkalinity in ILs led to increased solubility of acid gases (H_2S and CO_2) in the liquids. Due to the higher acidity and polarity of H_2S molecules compared to CO_2, the solubility of H_2S gas in IL was higher than that of CO_2.

Acknowledgements

This research project was financially supported by Ferdowsi University of Mashhad, Iran (grant No. 3/48448). The computations were partly carried out in the

High-Performance Computing (HPC) Center at Ferdowsi University of Mashhad. The author would like to appreciate the HPC cooperation.

Conflict of interest

There is no conflict of interest.

Author details

Fatemeh Moosavi
Faculty of Science, Department of Chemistry, Ferdowsi University of Mashhad, Mashhad, Iran

*Address all correspondence to: moosavibaigi@um.ac.ir

References

[1] Sazanova TS, Akhmetshina AI, Petukhov AN, Vorotyntsev AV, Suvorov SS, Barysheva AV, et al. The cation effect on the free volume and the solubility of H_2S and CO_2 in ionic liquids based on Bis(2-ethylhexyl) sulfosuccinate anion. Membranes. 2023;**13**(2):238

[2] Afsharpour A, Haghtalab A. Simultaneous measurement absorption of CO_2 and H_2S mixture into aqueous solutions containing diisopropanolamine blended with 1-butyl-3-methylimidazolium acetate ionic liquid. International Journal of Greenhouse Gas Control. 2017;**58**:71-80

[3] Hussain SMS, Adewunmi AA, Alade OS, Murtaza M, Mahboob A, Khan HJ, et al. A review of ionic liquids: Recent synthetic advances and oilfield applications. Journal of the Taiwan Institute of Chemical Engineers. 2023;**153**:105195

[4] Budzianowski WM. Single solvents, solvent blends, and advanced solvent systems in CO_2 capture by absorption: A review. International Journal of Global Warming. 2015;7(2):184-225

[5] Dutcher B, Fan M, Russell AG. Amine-based CO_2 capture technology development from the beginning of 2013 - A review. ACS Applied Materials & Interfaces. 2015;**7**(4):2137-2148

[6] Yamada H. Amine-based capture of CO_2 for utilization and storage. Polymer Journal. 2021;**53**(1):93-102

[7] Julio AAV, Castro-Amoedo R, Maréchal F, González AM, Escobar Palacio JC. Exergy and economic analysis of the trade-off for design of post-combustion CO_2 capture plant by chemical absorption with MEA. Energy. 2023;**280**:128004

[8] Kopitha K, Elakneswaran Y, Kitagaki R, Saito R, Tsujino M, Nishida A, et al. N-methyldiethanolamine (MDEA) as an effective CO_2 absorbent for direct air capture (DAC) in cement-based materials. Chemical Engineering Journal. 2023;**475**:146067

[9] Li F, Hemmati A, Rashidi H. Industrial CO_2 absorption into methyldiethanolamine/piperazine in place of monoethanolamine in the absorption column. Process Safety and Environmental Protection. 2020;**142**:83-91

[10] Shang D, Liu X, Bai L, Zeng S, Xu Q, Gao H, et al. Ionic liquids in gas separation processing. Current Opinion in Green and Sustainable Chemistry. 2017;**5**:74-81

[11] Frias RM, Follegatti-Romero LM, Follegatti-Romero LA. Modelling mixtures of ammonium and pyridinium-based ionic liquids and carbon dioxide with the Cubic Plus Association Equation of state. Fluid Phase Equilibria. 2023;**574**:113910

[12] Farzaneh A, Saghatoleslami N, Feyzi Y. Transient H_2S content rise in the effluent of a natural gas treating unit: Role of COS hydrolysis and heat stable amine salts. Process Safety and Environmental Protection. 2021;**153**:84-93

[13] Mazzetti MJ, Skagestad R, Mathisen A, Eldrup NH. CO_2 from natural gas sweetening to kick-start EOR in the north sea. Energy Procedia. 2014;**63**:7280-7289

[14] Ochędzan-Siodłak W, Dziubek K, Siodłak D. Densities and viscosities of imidazolium and pyridinium chloroaluminate ionic liquids. Journal of Molecular Liquids. 2013;**177**:85-93

[15] Singh SK, Savoy AW. Ionic liquids synthesis and applications: An overview. Journal of Molecular Liquids. 2020;**297**:112038

[16] Hallett JP, Welton T. Room-temperature ionic liquids: Solvents for synthesis and catalysis. 2. Chemical Reviews. 2011;**111**(5):3508-3576

[17] Moosavi F. The structure of supported ionic liquids at the interface. In: Kadokawa J-i, editor. Ionic Liquids-New Aspects for the Future. London, UK: IntechOpen; 2013

[18] Poole CF, Poole SK. Extraction of organic compounds with room temperature ionic liquids. Journal of Chromatography A. 2010;**1217**(16):2268-2286

[19] Babamohammadi S, Shamiri A, Aroua MK. A review of CO_2 capture by absorption in ionic liquid-based solvents. Reviews in Chemical Engineering. 2015;**31**(4):383-412

[20] Ren S, Hou Y, Zhang K, Wu W. Ionic liquids: Functionalization and absorption of SO_2. Green Energy & Environment. 2018;**3**(3):179-190

[21] Flieger J, Flieger M. Ionic liquids toxicity-benefits and threats. International Journal of Molecular Sciences. 2020;**21**(17):1-41

[22] Siami H, Razmkhah M, Moosavi F. Cation functional group effect on SO_2 absorption in amino acid ionic liquids. Frontiers in Chemistry. 2023;**11**:1-12

[23] Siami H, Razmkhah M, Moosavi F. Does side chain group of anion affect absorption of SO_2 in amino acid ionic liquid? Journal of Molecular Liquids. 2023;**376**:121479

[24] Taji K, Moosavi F. Molecular simulation of CO_2 capturing by dual functionalized phosphonium-based amino acid ionic liquids. Materials Today Communications. 2023;**37**:106945

[25] Ghatee MH, Moosavi F, Zolghadr AR, Jahromi R. Critical-point temperature of ionic liquids from surface tension at liquid-vapor equilibrium and the correlation with the interaction energy. Industrial and Engineering Chemistry Research. 2010;**49**(24):12696-12701

[26] Ghatee MH, Zare M, Moosavi F, Zolghadr AR. Temperature-dependent density and viscosity of the ionic liquids 1-alkyl-3-methylimidazolium iodides: Experiment and molecular dynamics simulation. Journal of Chemical and Engineering Data. 2010;**55**(9):3084-3088

[27] Ghatee MH, Zare M, Zolghadr AR, Moosavi F. Temperature dependence of viscosity and relation with the surface tension of ionic liquids. Fluid Phase Equilibria. 2010;**291**(2):188-194

[28] Ghatee MH, Moosavi F. Physisorption of hydrophobic and hydrophilic 1-alkyl-3-methylimidazolium ionic liquids on the graphenes. Journal of Physical Chemistry C. 2011;**115**(13):5626-5636

[29] Ghatee MH, Zolghadr AR, Moosavi F, Ansari Y. Studies of structural, dynamical, and interfacial properties of 1-alkyl-3-methylimidazolium iodide ionic liquids by molecular dynamics simulation. Journal of Chemical Physics. 22012;**136**(12):124706(1-14)

[30] Moosavi F, Razmkhah M. Structural view of hydrophobic ionic liquid on graphene: Comparing static and ab initio computer simulations. ECS Journal of Solid State Science and Technology. 2015;**4**(12):M77-M87

[31] Sadeghi Moghadam B, Razmkhah M, Hamed Mosavian MT, Moosavi F. Molecular dynamics simulation of amino acid ionic liquids near a graphene electrode: Effects of alkyl side-chain length. Physical Chemistry Chemical Physics. 2016;**18**(48):33053-33067

[32] Ebrahimi M, Moosavi F. The effects of temperature, alkyl chain length, and anion type on thermophysical properties of the imidazolium based amino acid ionic liquids. Journal of Molecular Liquids. 2018;**250**:121-130

[33] Razmkhah M, Hamed Mosavian MT, Moosavi F. Structural analysis of an amino acid ionic liquid: Bulk and electrical double layer. Journal of Molecular Liquids. 2018;**268**:506-516

[34] Razmkhah M, Hamed Mosavian MT, Moosavi F. What is the effect of polar and nonpolar side chain group on bulk and electrical double layer properties of amino acid ionic liquids? Electrochimica Acta. 2018;**285**:393-404

[35] Nadimi H, Housaindokht MR, Moosavi F. The effect of anion on aggregation of amino acid ionic liquid: Atomistic simulation. Journal of Molecular Graphics and Modelling. 2020;**101**:107733

[36] Wang J, Zeng S, Bai L, Gao H, Zhang X, Zhang S. Novel ether-functionalized pyridinium chloride ionic liquids for efficient SO_2 capture. Industrial & Engineering Chemistry Research. 2014;**53**(43):16832-16839

[37] Wang X, Zeng S, Wang J, Shang D, Zhang X, Liu J, et al. Selective separation of hydrogen sulfide with pyridinium-based ionic liquids. Industrial & Engineering Chemistry Research. 2018;**57**(4):1284-1293

[38] Duan H, Liu F, Hussain S, Dong H, Zhang X, Cheng Z. Degradation processes of protic ionic liquids for NH_3 separation. New Journal of Chemistry. 2023;**47**(23):11275-11284

[39] Kapoor U, Banerjee A, Shah JK. Evaluation of the predictive capability of ionic liquid force fields for CH_4, CO_2, NH_3, and SO_2 phase equilibria. Fluid Phase Equilibria. 2019;**492**:161-173

[40] Yuan L, Gao H, Jiang H, Zeng S, Li T, Ren B, et al. Experimental and thermodynamic analysis of NH_3 absorption in dual-functionalized pyridinium-based ionic liquids. Journal of Molecular Liquids. 2021;**323**:114601

[41] Li P, Shang D, Tu W, Zeng S, Nie Y, Bai L, et al. NH_3 absorption performance and reversible absorption mechanisms of protic ionic liquids with six-membered N-heterocyclic cations. Separation and Purification Technology. 2020;**248**:117087

[42] Yokozeki A, Shiflett MB. Ammonia solubilities in room-temperature ionic liquids. Industrial & Engineering Chemistry Research. 2007;**46**(5):1605-1610

[43] Yokozeki A, Shiflett MB. Vapor–liquid equilibria of ammonia+ionic liquid mixtures. Applied Energy. 2007;**84**(12):1258-1273

[44] Qiu R, Luo X, Yang L, Li J, Chen X, Peng C, et al. Regulated threshold pressure of reversibly sigmoidal NH_3 absorption isotherm with ionic liquids. ACS Sustainable Chemistry & Engineering. 2020;**8**(3):1637-1643

[45] Wang L, Zhang Y, Liu Y, Xie H, Xu Y, Wei J. SO_2 absorption in pure ionic liquids: Solubility and functionalization. Journal of Hazardous Materials. 2020;**392**:122504

[46] Zeng S, Gao H, Zhang X, Dong H, Zhang X, Zhang S. Efficient and reversible capture of SO_2 by pyridinium-based ionic liquids. Chemical Engineering Journal. 2014;**251**:248-256

[47] Zeng S, He H, Gao H, Zhang X, Wang J, Huang Y, et al. Improving SO_2 capture by tuning functional groups on the cation of pyridinium-based ionic liquids. RSC Advances. 2015;**5**(4):2470-2478

[48] Hong SY, Im J, Palgunadi J, Lee SD, Lee JS, Kim HS, et al. Ether-functionalized ionic liquids as highly efficient SO_2 absorbents. Energy & Environmental Science. 2011;**4**(5):1802-1806

[49] Smith W, Forester TR. DL_POLY_2.0: A general-purpose parallel molecular dynamics simulation package. Journal of Molecular Graphics. 1996;**14**(3):136-141

[50] Humphrey W, Dalke A, Schulten K. VMD: Visual molecular dynamics. Journal of Molecular Graphics. 1996;**14**(1):33-38

[51] Bandrés I, Alcalde R, Lafuente C, Atilhan M, Aparicio S. On the viscosity of pyridinium based ionic liquids: An experimental and computational study. The Journal of Physical Chemistry B. 2011;**115**(43):12499-12513

[52] Shi W, Myers CR, Luebke DR, Steckel JA, Sorescu DC. Theoretical and experimental studies of CO_2 and H_2 separation using the 1-ethyl-3-methylimidazolium acetate ([emim][CH_3COO]) ionic liquid. The Journal of Physical Chemistry B. 2012;**116**(1):283-295

[53] Aparicio S, Atilhan M. A computational study on choline benzoate and choline salicylate ionic liquids in the pure state and after CO_2 adsorption. The Journal of Physical Chemistry B. 2012;**116**(30):9171-9185

[54] Aparicio S, Atilhan M. Nanoscopic vision on fuel dearomatization using ionic liquids: The case of piperazine-based fluids. Energy & Fuels. 2013;**27**(5):2515-2527

[55] Kamath G, Lubna N, Potoff JJ. Effect of partial charge parametrization on the fluid phase behavior of hydrogen sulfide. The Journal of Chemical Physics. 2005;**123**(12):124505(1-7)

[56] Martin MG, Siepmann JI. Transferable potentials for phase equilibria. 1. United-atom description of n-alkanes. The Journal of Physical Chemistry B. 1998;**102**(14):2569-2577

[57] Chandran A, Prakash K, Senapati S. Structure and dynamics of acetate anion-based ionic liquids from molecular dynamics study. Chemical Physics. 2010;**374**(1):46-54

[58] Cadena C, Zhao Q, Snurr RQ, Maginn EJ. Molecular modeling and experimental studies of the thermodynamic and transport properties of pyridinium-based ionic liquids. The Journal of Physical Chemistry B. 2006;**110**(6):2821-2832

[59] Larive CK, Lin M, Piersma BJ, Carper WR. Diffusion ordered spectroscopy of room temperature chloroaluminate melts. The Journal of Physical Chemistry. 1995;**99**(33):12409-12412

[60] Tsimpanogiannis IN, Moultos OA, Franco LFM, Spera MBM, Erdős M,

Economou IG. Self-diffusion coefficient of bulk and confined water: A critical review of classical molecular simulation studies. Molecular Simulation. 2019;**45**(4-5):425-453

[61] Morrow TI, Maginn EJ. Molecular dynamics study of the ionic liquid 1-n-butyl-3-methylimidazolium hexafluorophosphate. The Journal of Physical Chemistry B. 2002;**106**(49):12807-12813

[62] Morganti JD, Hoher K, Ribeiro MCC, Ando RA, Siqueira LJA. Molecular dynamics simulations of acidic gases at interface of quaternary ammonium ionic liquids. The Journal of Physical Chemistry C. 2014;**118**(38):22012-22020

[63] Kowsari MH, Alavi S, Ashrafizaadeh M, Najafi B. Molecular dynamics simulation of imidazolium-based ionic liquids. I. Dynamics and diffusion coefficient. The Journal of Chemical Physics. 2008;**129**(22):224508(1-13)

[64] Anthony JL, Anderson JL, Maginn EJ, Brennecke JF. Anion effects on gas solubility in ionic liquids. The Journal of Physical Chemistry B. 2005;**109**(13):6366-6374

[65] Bhargava BL, Balasubramanian S. Probing anion–carbon dioxide interactions in room temperature ionic liquids: Gas phase cluster calculations. Chemical Physics Letters. 2007;**444**(4):242-246

[66] Seki T, Grunwaldt J-D, Baiker A. In situ attenuated total reflection infrared spectroscopy of imidazolium-based room-temperature ionic liquids under "supercritical" CO_2. The Journal of Physical Chemistry B. 2009;**113**(1):114-122

[67] Ren S, Hou Y, Tian S, Chen X, Wu W. Correction to "what are functional ionic liquids for the absorption of acidic gases?". The Journal of Physical Chemistry B. 2013;**117**(17):5394-5395

[68] Lide DR. CRC Handbook of Chemistry and Physics. Vol. 85. Boca Raton: CRC Press LLC; 2004

[69] Wang L-y, Xu Y-l, Li Z-d, Wei Y-n, Wei J-p. CO_2/CH_4 and H_2S/CO_2 selectivity by ionic liquids in natural gas sweetening. Energy & Fuels. 2018;**32**(1):10-23

[70] Bara JE, Gabriel CJ, Lessmann S, Carlisle TK, Finotello A, Gin DL, et al. Enhanced CO_2 separation selectivity in oligo(ethylene glycol) functionalized room-temperature ionic liquids. Industrial & Engineering Chemistry Research. 2007;**46**(16):5380-5386

[71] Brooks BR, Bruccoleri RE, Olafson BD, States DJ, Swaminathan S, Karplus M. CHARMM: A program for macromolecular energy, minimization, and dynamics calculations. Journal of Computational Chemistry. 1983;**4**(2):187-217

Chapter 8

Perspective Chapter: Introduction to Hydrogels – Definition, Classifications, Applications and Methods of Preparation

Behrooz Maleki, Pouya Ghamari Kargar, Samaneh Sedigh Ashrafi and Milad Ghani

Abstract

Hydrogel products are a group of polymeric materials that possess a hydrophilic structure, allowing them to retain significant amounts of water within their three-dimensional networks. The development of new materials is crucial for advancing technologies, and this often involves the innovative combination of existing components. A combination that incorporates both a polymer hydrogel network and nanoparticles can be achieved by combining metals, non-metals, metal oxides, and polymeric moieties. The composite material's functionality will be enhanced by this amalgamation, which has applications in various fields such as catalysis, electronics, biosensing, drug delivery, nano-medicine, and environmental remediation. The incorporation of nanoparticles into hydrogels can result in synergistic property enhancements, such as improved mechanical strength of the hydrogel and a reduction in nanoparticle aggregation. These mutually beneficial effects have attracted significant interest from multidisciplinary research groups over the past decade. In this chapter, we delve into recent advancements in nanoparticle-hydrogel composites, focusing on their synthesis, design, potential applications, and the inherent challenges associated with these exciting materials.

Keywords: hydrogel, hydrogel concept, hydrogel classification, hydrogel application, ionic liquid

1. Introduction

Hydrogels are unique materials that consist of a three-dimensional network of hydrophilic polymers. They can absorb and retain large amounts of water or aqueous fluids without completely dissolving. The water-retaining capacity of hydrogels is attributed to the functional groups attached to the polymer backbone, such as hydroxylic, amidic, carboxylic, sulfonic, and primary amidic groups. These groups have a high affinity for water. The cross-links between the polymer chains provide structural integrity to the hydrogel and prevent its disintegration. Hydrogels can exhibit softness, high efficiency,

 IntechOpen

and the ability to store materials, making them versatile materials for various applications. Hydrogels also act as matrices that allow the diffusion of certain solute molecules while holding water together. They form semi-open macromolecular network systems with intertwined or cross-linked chains of different lengths. When exposed to a compatible solvent like water or biological fluids, hydrogels swell and trap a large amount of solvent molecules within their pores. Hydrogels can undergo phase transitions, including gel-sol transitions, in response to physical or chemical stimuli. Physical stimuli such as temperature, pressure, magnetic and electric fields, light intensity, and solvent compositions can trigger these transitions. Chemical or biochemical stimuli include ions, pH changes, and specific chemical compositions. The response of hydrogels to external stimuli depends on factors such as charge density, degree of cross-linkage, presence of free chains, and the nature of the monomer used.

The synthesis of hydrogels involves a series of chemical events, including polymerization and cross-linking of multifunctional monomers. One method is the reaction

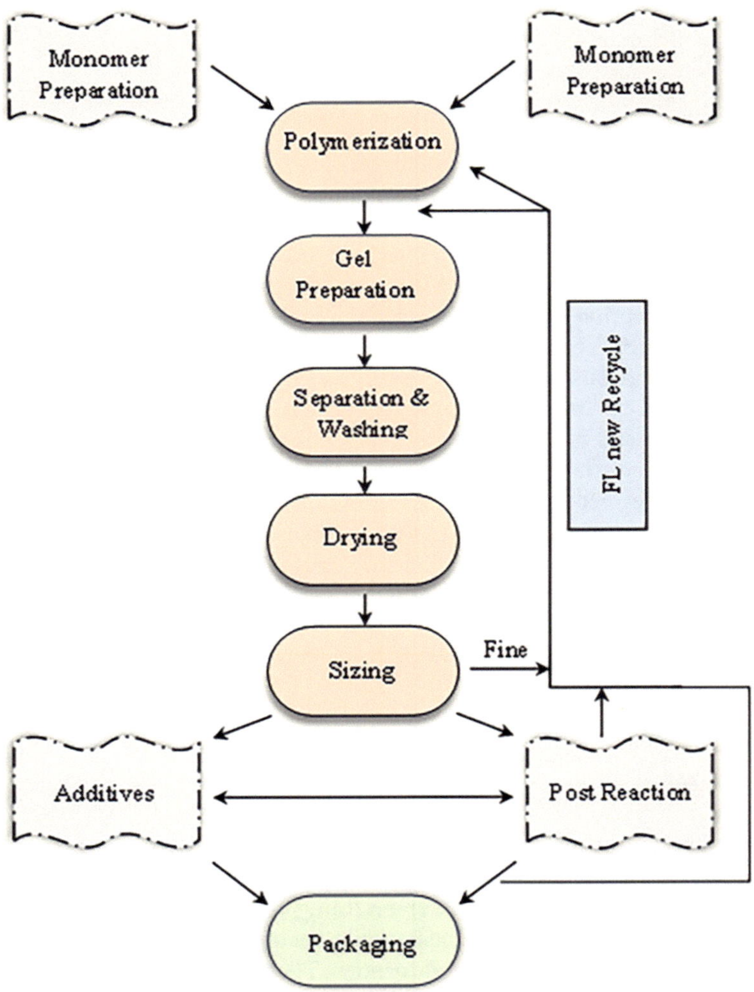

Figure 1.
Hydrogel preparation block diagram (solution polymerization/cross-linking procedure).

between polymers and cross-linking agents. Another approach is a multistep process where a polymer with reactive groups is synthesized first, followed by cross-linking with suitable agents. These reactions can involve various chemical interactions, such as ionic forces, hydrogen bonds, van der Waals interactions, primary covalent cross-linkages, affinity interactions, and hydrophobic interactions. These interactions contribute to holding the molecules within the three-dimensional matrix of the hydrogel. Hydrogels can be synthesized from both natural and synthetic polymers. Natural polymers such as agarose, alginate, collagen, chitosan, chondroitin sulfate, dextran, fibrin, gelatin, hyaluronic acid, and pectin can be used. Synthetic polymers include polyacrylic acid, polyacrylamide, polyvinyl alcohol, polyethylene oxide, polyhydroxyethyl methacrylate, poly-2-hydroxyethyl methacrylate (PHEMA), polypropylene fumarate-co-ethylene glycol, polypeptides, and poly(N-isopropylacrylamide). In many hydrogel formulations, a combination of natural and synthetic polymers is used to enhance the physicochemical properties and biocompatibility. This can be achieved through interpenetrating polymer networks (IPN) or semi-interpenetrating polymer networks (semi-IPN).

According to the findings of Javad Alaei et al. [1], the industrial production of hydrogels encompasses solution polymerization, reversed suspension polymerization, and reversed emulsion polymerization techniques. **Figure 1** presents a schematic diagram illustrating the key steps involved in the manufacturing of hydrogels at semi-pilot and industrial scales.

The use of classical organic reactions between the functional groups of the polymer being utilized is common for cross-linking polymeric materials. The Huisgen cycloaddition, the Diels-Alder reaction, the Michael addition, and the Schiff base reaction are just a few of the reactions that are notable (**Figure 2**) [2].

Indeed, hydrogels have gained significant attention in recent years due to their unique characteristics and versatile properties. They find applications in various industries, particularly in the biomedical field. Here are some key areas where hydrogels are extensively used: biomedical engineering, 3D printing, agriculture, food packaging, separation technology, cosmetics and personal care, biosensors, and bio-microelectromechanical systems (Bio-MEMS). The unique characteristics of hydrogels, including their biocompatibility, water retention capacity, and tunable properties, make them highly attractive for a wide range of applications across multiple industries. Ongoing research and development in the field of hydrogels continues to expand their potential applications and impact.

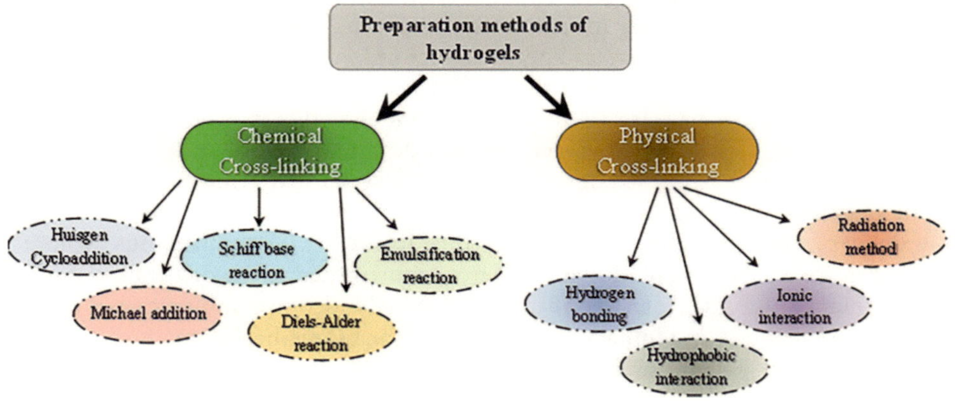

Figure 2.
Schematic overview of the different cross-linking methods for preparing hydrogels.

2. History of hydrogels

Bacterial biofilms and plant structures have long been observed as water-swollen motifs in nature. Throughout human history, gelatin and agar were discovered and utilized for various purposes. However, the specific history of hydrogels as a distinct group of materials designed for biological applications can be traced back to more recent times. The term "hydrogel" was coined by Lee, Kwon, and Park in 1894. However, the substance they described was not a hydrogel in the modern sense but rather an inorganic salt-based gel. Despite this, the term "hydrogel" has endured and continues to be used to describe a wide range of water-swollen materials. In the nineteenth century, Thomas Graham conducted the first scientific study of gels. He produced a silica gel using sol–gel chemistry, laying the groundwork for the understanding of gel formation. In 1974, Paul Flory was awarded the Nobel Prize in Chemistry for his contributions to polymer science, including the development of macromolecules and polymers. His work played a pivotal role in advancing the field of polymer science. Another notable figure in the history of polymers is Alan G. Wardhaugh Treloar, who made significant contributions to the study of polymer elasticity. His work greatly influenced the understanding of polymer behavior and properties. The combined efforts of these pioneers and many others have shaped the field of hydrogels and polymer science as we know it today. Their research and discoveries have paved the way for the development and application of hydrogels in various fields, including biomedicine, materials science, and engineering [3, 4].

Hydrogels have indeed played a significant role in biomedical applications and have been used with human tissues as biocompatible materials. The development of the first synthetic hydrogel based on poly-2-hydroxyethyl methacrylate (PHEMA) for contact lens applications was described by Wichterle and Lim [5]. This breakthrough led to the use of PHEMA hydrogels as contact lenses, marking one of the early successful applications of hydrogels in 1960. This development paved the way for the emergence of biomedical hydrogels, which are specifically designed for applications within the human body. Since the 1970s, numerous studies have been conducted to explore the diverse biomedical applications of hydrogels. In the 1980s, countries like Germany and France started incorporating hydrogels into diaper products, which further propelled advancements in the field of hydrogel research. As interest in biopolymers and hydrogels grew, more researchers began focusing on developing efficient and biocompatible hydrogels. This involved studying the properties and characteristics of hydrogel and polymer materials in-depth. The goal was to enhance the functionality, biocompatibility, and performance of hydrogels for various biomedical applications. Over time, advancements in materials science, polymer chemistry, and fabrication techniques have led to the development of a wide range of hydrogels with tailored properties. These hydrogels can now be designed to possess desired characteristics such as mechanical strength, controlled drug release, stimuli-responsiveness, and biodegradability, making them suitable for applications in tissue engineering, drug delivery, wound healing, and other biomedical fields. The continuous exploration and research in the field of hydrogels have contributed to the development of more efficient and versatile materials with enhanced biocompatibility, expanding the potential applications and impact of hydrogels in the biomedical domain.

There are three generations of hydrogel development: first, second, and third [6]. To obtain hydrogels with desirable swelling, physicochemical, and mechanical properties, first-generation hydrogels employ diverse cross-linking methods and modifications of monomers or polymers.

Second-generation hydrogels emerged in the mid-1970s, with a focus on incorporating responsive behavior to environmental stimuli. These hydrogels can respond to stimuli such as temperature, pH, solvent composition, molecule concentration, and ionic charge changes. Their responsive nature finds applications in drug delivery, in situ pore formation, polymerization, and cross-linking.

The current research is directed toward developing third-generation hydrogels, often referred to as smart hydrogels. These advanced hydrogels exhibit dynamic properties and can change their composition and texture in response to external stimuli like temperature, pH, solvent composition, and electric fields. They are considered smart materials and are being explored for a wide range of applications.

Over the past 50 years, hydrogels have found extensive applications in various sectors, including agriculture, biomaterials, pharmaceutics, biotechnology, and wastewater treatment. In recent years, hydrogels have made significant breakthroughs in tissue engineering, enabling advancements in wound healing, tissue repair, and regenerative medicine. The use of hydrogels in tissue engineering gained momentum with Lim and Sun's work in 1980, demonstrating the feasibility of using calcium alginate microcapsules for cell encapsulation. Yannas and colleagues made significant progress in the field in the 1980s by creating hydrogels made from natural polymers like collagen and shark cartilage to be used as artificial burn dressings. Hydrogels that are composed of natural and synthetic polymers have become popular for encapsulating cells and have been utilized in tissue engineering for healing and regenerating various tissues and organs. The field of hydrogels is continuously evolving, and the development of smart hydrogels with adjustable characteristics and responsive behavior opens up vast possibilities in engineering and medical applications. The potential for further advancements and applications in this field appears to be limitless [7].

3. Classifications of hydrogels

Hydrogels can be categorized based on various characteristics and factors (**Figure 3**) [8, 9]. Some common classification criteria include:

1. Physical Characteristics: Hydrogels can be classified based on physical mechanisms that contribute to their structure and properties, such as hydrophobic association, chain aggregation, crystallization, polymer chain conformation, and hydrogen bonding.

2. Swelling Characteristics: Hydrogels can be classified based on their swelling behavior, including the extent of swelling, response to stimuli, and the ability to retain or release water or other substances.

3. Manufacturing Process: Hydrogels can be classified based on the method of fabrication, such as physical gelation, chemical cross-linking, or a combination of both.

4. Origin or Source: Hydrogels can be classified based on their origin, distinguishing between natural hydrogels derived from biological sources (e.g., proteins, polysaccharides) and synthetic hydrogels produced from synthetic polymers.

5. Ionic Charge: Hydrogels can be classified based on their ionic charge, which can be positively charged (cationic), negatively charged (anionic), or neutral.

Figure 3.
Hydrogel classification: A diagrammatic depiction.

6. Biodegradation Rate: Hydrogels can be categorized based on their rate of biodegradation, distinguishing between fast-degrading hydrogels that break down relatively quickly and slow-degrading hydrogels that maintain their structure for a longer duration.

7. Cross-Linking: Hydrogels can be classified based on the type of cross-linking present, such as physical cross-linking (reversible interactions) or chemical cross-linking (covalent bonds). Chemical hydrogels have permanent and irreversible cross-linking, while physical hydrogels exhibit reversible conformational changes.

Additionally, dual-network hydrogels are a type of hydrogel that combines both physical and chemical cross-linking, enabling electrostatic interactions between the networks. This type of hydrogel overcomes some limitations associated with using solely physical or chemical hydrogels, providing high liquid absorption capacity over a wide pH range and increased sensitivity to pH changes compared to chemical hydrogels. Hydrogels can have diverse compositions and properties, and their classification can be based on a combination of factors such as physical condition, polymeric composition, pore size, physical appearance, and configuration. This classification helps in understanding and categorizing the wide range of hydrogel types available in both natural and synthetic forms.

3.1 Classification based on origin

Hydrogels can be classified into three main types based on their composition: natural hydrogels, synthetic hydrogels, and hybrid hydrogels:

1. Natural Hydrogels: These hydrogels are derived from natural polymers, such as polysaccharides and proteins. Examples of natural hydrogels include agarose, chitin, hyaluronic acid (HA), collagen, dextran, and gelatin. Natural hydrogels are biocompatible, biodegradable, and often possess high cell adhesion properties. They are synthesized from naturally occurring biomaterials and find applications in various biomedical fields.

2. Synthetic Hydrogels: Synthetic hydrogels are created through polymerization processes using artificial monomers. They are engineered materials with a wide range of chemical and mechanical properties. Examples of synthetic hydrogels include polyethylene glycol (PEG)-polylactic acid (PLA) hydrogels. Synthetic hydrogels are more inert compared to natural hydrogels and offer advantages such as high compatibility, low immunogenicity, and nontoxicity. They are commonly used in biomedical applications due to their controlled and tailored properties.

3. Hybrid Hydrogels: Hybrid hydrogels are a combination of natural hydrogels and synthetic polymer hydrogels. These hydrogels are designed to leverage the benefits of both natural and synthetic components. Natural polymers like chitosan, collagen, and dextran are mixed with artificial polymers such as polyvinyl alcohol and poly(N-isopropylacrylamide) to create hybrid hydrogels. By combining different materials, hybrid hydrogels can exhibit enhanced properties and functionalities, making them suitable for various biomedical applications.

The classification of hydrogels into natural, synthetic, and hybrid types allows researchers and engineers to choose the most appropriate hydrogel materials based on their specific requirements and desired properties.

3.2 Classification based on polymeric composition

Hydrogels can be further classified into different types based on the composition and structure of their polymer networks. Some of these types include:

1. Homopolymer Hydrogels: Homopolymeric hydrogels are cross-linked polymer networks formed from a single type of monomer. The choice of monomer, polymerization method, and cross-linker influences the structural composition of these hydrogels. For example, polyethylene glycol (PEG) homopolymer hydrogels are often used for controlled release applications in drug delivery systems due to their biocompatibility and ability to control the release of medicines, proteins, and biomolecules.

2. Copolymer Hydrogels: Copolymeric hydrogels consist of two or more types of monomer species in their polymeric network. Typically, one of the monomers is hydrophilic, enabling the hydrogel to swell. Copolymer hydrogels can have various arrangements of monomers, including graft, random, block, and alternate configurations. The hydrogels are formed by polymerizing or cross-linking two monomers using a single initiator and cross-linker. The physical cross-linking in copolymer hydrogels can occur through interactions such as hydrogen bonding, chain aggregation, ion-polymer complexation, and ionic interactions. Examples of copolymer hydrogels include carboxymethyl cellulose and carboxymethyl chitosan-based hydrogels used for metal ion adsorption.

3. Multipolymer Hydrogels: Multipolymer hydrogels are formed from three or more monomers through polymerization and cross-linking reactions. The independent polymer components can be of synthetic or natural origin and are cross-linked together to form a network. An example of a multipolymer hydrogel is poly (acrylic acid-2-hydroxy ethyl methacrylate) or gelatin hydrogel.

4. Interpenetrating Network (IPN): Interpenetrating network hydrogels consist of two interlaced polymer networks without chemical interactions between them. The first polymer network is linear, while the second polymer network is cross-linked. The linear network of the first polymer diffuses into the second polymer, resulting in interlocking and strong cross-linking between the two networks. IPN hydrogels can overcome thermodynamic incompatibility and restricted phase separation. Semi-interpenetrating networks (Semi-IPN) are a type of IPN where the linear polymers are scattered inside the first network. IPN hydrogels can be synthesized through simultaneous or sequential methods.

In summary, hydrogels can be classified into various types based on their composition and structure, including homopolymer hydrogels, copolymer hydrogels, multipolymer hydrogels, and interpenetrating network (IPN) hydrogels. These classifications help in understanding the different types of hydrogels and their applications in various fields, such as drug delivery, tissue engineering, and biomaterials.

3.3 Classification based on ionic charge

Hydrogels can also be classified based on the electric charge on their cross-linked chains. Here are the three types:

1. Neutral (Nonionic) Hydrogels: Neutral hydrogels have no net charge on their side groups or backbone. They do not contain any ionizable groups. These hydrogels are electrically neutral and do not exhibit specific interactions with ions or charged molecules. They are typically used in applications where a neutral environment is desired, such as in contact lenses.

2. Ionic Hydrogels: Ionic hydrogels contain ionizable groups that can be either cationic or anionic. These groups introduce a net positive or negative charge to the hydrogel. Cationic hydrogels have positively charged groups, such as amines, while anionic hydrogels have negatively charged groups, such as carboxylic acids or sulfonic acids. The swelling properties of ionic hydrogels can be pH-dependent. Cationic hydrogels tend to swell more at lower pH values, while anionic hydrogels show increased swelling at higher pH values.

3. Ampholytic (Amphoteric) Hydrogels: Ampholytic hydrogels contain both negative (basic) and positive (acidic) charges on the same polymer chain. These charges can coexist and are balanced at the isoelectric point. Ampholytic hydrogels can exhibit different properties depending on the pH of the surrounding environment. They can respond to changes in pH by altering their swelling behavior or charge distribution. These hydrogels find applications in pH-sensitive drug delivery systems and biomaterials.

Additionally, there is a category known as zwitterionic hydrogels, where each structural unit contains both anionic and cationic groups. This balanced charge distribution provides unique properties to zwitterionic hydrogels, such as excellent biocompatibility, low fouling, and resistance to protein adsorption. Zwitterionic hydrogels have gained attention for applications in biomedicine, such as coatings for medical devices and tissue engineering scaffolds. The classification of hydrogels based on electric charge helps in understanding their behavior and properties in different environments, and it enables the design of hydrogels with specific functionalities for various applications.

3.4 Classification based on configuration

Hydrogels can also be classified based on their structural organization, specifically regarding the presence or absence of crystalline regions. Here are the three types:

1. Amorphous (Noncrystalline) Hydrogels: Amorphous hydrogels have randomly arranged macromolecular chains in their polymeric networks. They lack a well-defined crystalline structure. The absence of crystalline regions gives amorphous hydrogels a more flexible and disordered nature.

2. Semicrystalline Hydrogels: Semicrystalline hydrogels exhibit a complex blend of amorphous and crystalline phases. In these hydrogels, there are sections or domains where the macromolecular chains are organized into a more ordered, crystalline structure. These crystalline regions coexist with the amorphous regions, resulting in a combination of mechanical properties associated with both structures.

3. Crystalline Hydrogels: Crystalline hydrogels consist primarily of a well-defined crystalline phase. The macromolecular chains within the hydrogel are highly organized and aligned in a repeating, three-dimensional lattice structure. These hydrogels have distinct crystalline regions and exhibit unique mechanical, thermal, and chemical properties associated with crystallinity.

The structural organization of hydrogels, whether they are amorphous, semicrystalline, or crystalline, can significantly affect their mechanical strength, swelling behavior, and other properties. The type of structural organization desired in a hydrogel depends on the specific application and the desired properties required for that application.

3.5 Classification based on cross-linking

Hydrogels can be classified into two types based on the nature of cross-linking: physical hydrogels and chemical hydrogels.

1. Physical Hydrogels: Physical hydrogels are cross-linked through reversible physical interactions rather than covalent chemical bonds. These physical interactions can include crystallization, hydrogen bonding, hydrophobic interactions, and electrostatic interactions. Physical hydrogels can undergo reversible changes in their structure and properties in response to external stimuli, such as temperature, pH, or solvent composition. These hydrogels can exhibit tunable gelation and sol–gel transitions, making them suitable for applications such as drug delivery, tissue engineering, and sensors.

2. Chemical Hydrogels: Chemical hydrogels are cross-linked through covalent chemical bonds, resulting in the formation of permanent junctions within the polymer network. The cross-linking can occur through various chemical reactions, such as polymerization, condensation, or click chemistry reactions. Chemical hydrogels have stable and permanent cross-links, providing them with robust mechanical properties and resistance to changes in environmental conditions. These hydrogels are commonly used in applications requiring long-term stability and structural integrity, such as biomedical implants, wound dressings, and scaffolds for tissue engineering.

The classification of hydrogels into physical and chemical types is based on the underlying cross-linking mechanism. This classification is essential for understanding the behavior and properties of hydrogels and selecting the appropriate type for specific applications.

4. Applications of hydrogel

In the past few decades, hydrogels due to their unique properties including abundance, cheapness, renewableness, non-toxicity, biodegradability, and biocompatibility have been used in various industries such as food, packaging, pharmaceuticals, agriculture, biomedical and bioengineering applications, in the construction of technical and electronic devices, also as absorbents to remove pollutants in environmental applications, and in ionic liquids. Considering the importance and diverse capabilities of these compounds as promising materials in various applications, in this section, their important applications in various fields have been reviewed.

4.1 Food packaging

The hydrogel is comprised of a complex network structure made up of polymeric chains, creating a porous, three-dimensional (3D) network. The formation of such a structure is achieved by cross-linking polymer chains of natural or synthetic polymers that have the capability to absorb and retain a significant amount of water in their interstitial frameworks (**Figure 4**).

In recent years, the use of sustainable packaging resources in the circular economy framework has gained considerable attention as a way to decrease waste and decrease the negative environmental impact of packaging materials. Bio-based hydrogels are being examined for their potential applications in various fields, including food packaging, in line with these trends. The hydrophilic nature of hydrogels is a promising solution for food packaging systems, especially in controlling moisture levels and transporting bioactive substances, which can significantly impact the shelf life of food products. In general, the creation of cellulose-based hydrogels (CBHs) and their offshoots has resulted in hydrogels that are appealing due to their flexibility, water absorption, swelling capacity, biocompatibility, biodegradability, stimulus sensitivity, and cost-effectiveness. The purpose of this review is to provide an overview of the latest trends and applications of CBHs in the food packaging industry, with a focus on CBH sources, processing methods, and cross-linking methods used to create hydrogels through physical, chemical, and polymerization processes [10]. Chitosan (CH), which is the only natural cationic polysaccharide, has multiple bioactive properties. Moreover, chitosan was the ideal biomaterial for preparing hydrogels due

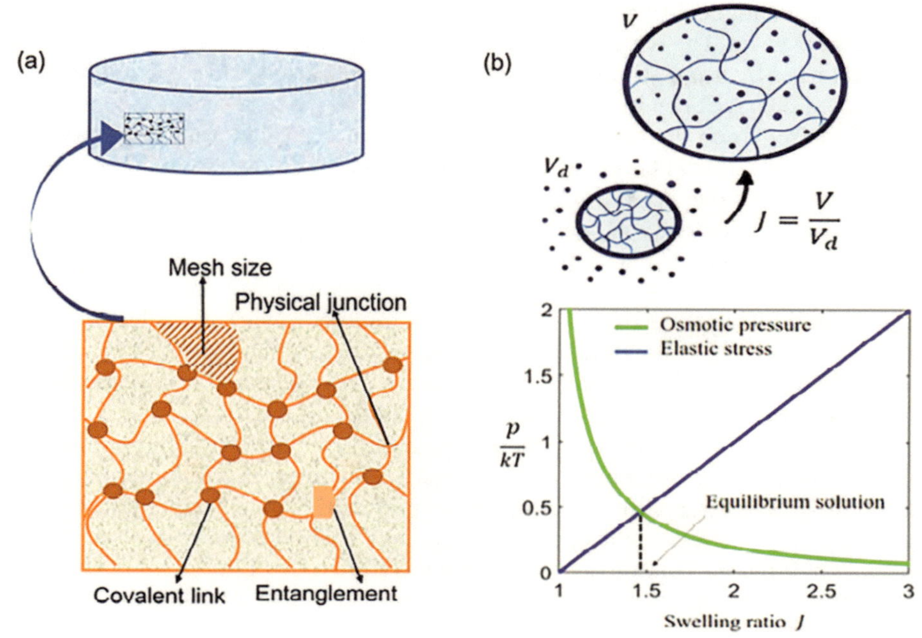

Figure 4.
Structural chemistry of a hydrogel: (a) the structure of a hydrogel at the molecular level is the subject of structural chemistry. (b) the relationship between the swelling ratio of the hydrogel and osmotic pressure and elastic stress.

to its desirable biocompatibility and biodegradability. CH-based hydrogels have been extensively studied for their applications in delivery systems and tissue engineering. Additionally, they have shown great potential and garnered considerable attention in recent years for use in biosensors and packaging materials [11]. Sugarcane bagasse's nanocellulose has been used to develop a hydrogel that can be used as a colorimetric indicator for monitoring the freshness of chicken breasts. The preparation of nanocellulose from sugarcane bagasse cellulose filaments was done through TEMPO-mediated oxidation, and then it was utilized to create a resilient hydrogel matrix by cross-linking with Zn^{2+} [12]. This review examines the current status of natural hydrogels for sustainable food packaging. The hydrogels are based on proteins, polysaccharides, lipids, and other biodegradable polymers. The review also focuses on the functionalization of these hydrogels with bioactive compounds, particularly phenolic compounds. Incorporating them into the hydrogel's formulation appears to enhance the mechanical, antioxidant, and antimicrobial properties of the resulting films, as well as extend the shelf life of food products. This review compiles recent applications of natural hydrogels in food. Although current research is still in its early stages, the results obtained thus far are promising. This represents an innovative, sustainable alternative to plastic materials for extending the shelf life of food products [13].

4.2 Pharmaceuticals

The versatility and potential demand of hydrogels can be attributed to their ability to absorb large amounts of aqueous solutions without losing shape or mechanical properties, as well as their ability to load drugs of different types, including hydrophobic and biomolecules. Many pathways of synthesis have been developed, particularly for chemical/permanent hydrogels, as they have been explored in numerous studies for years. The regulation of properties such as targeting and drug release can be enhanced by using stimuli-responsive hydrogels, also known as intelligent materials, which have been investigated. The control of particle size has led to similar studies of hydrogels on the micro- and nanoscale, which have increased the possibilities for applications of so-called XXI century materials even more. Our objective was to provide an overview of recent studies on the synthesis, biomedical, and pharmaceutical applications of macro-, micro-, and nanogels. The size of the particles obtained can determine the classification of hydrogels as macro-, micro-, or nanogels. When they are smaller than 100 nm, they are usually considered nanogels, while gels with particle sizes bigger than this (up to the micrometer range) are called microgels. Finally, if these gels have particle sizes bigger than 100 μm, they are usually called macrogels (**Figure 5**) [14]. Cellulose and lignin are natural, abundant, biodegradable, and environmentally friendly polymers. In addition, the presence of reactive functional groups and the ability to chemically modify cellulose and lignin provide great potential for development of products for the medical sciences and pharmaceuticals. Hydrogels in drug delivery systems have potential to give better performance of drugs. The properties of hydrogels and their constituent polymers are among the most challenging issues in this field. Research has shown that biopolymers, mainly cellulose and lignin, are excellent components for drug delivery hydrogels. These two biopolymers and their derivatives can be used as matrix hydrogels and cross-linking agents, improving mechanical properties and designing an engineered function for hydrogels, including their usage as pharmaceutical agents. The use of cellulose and

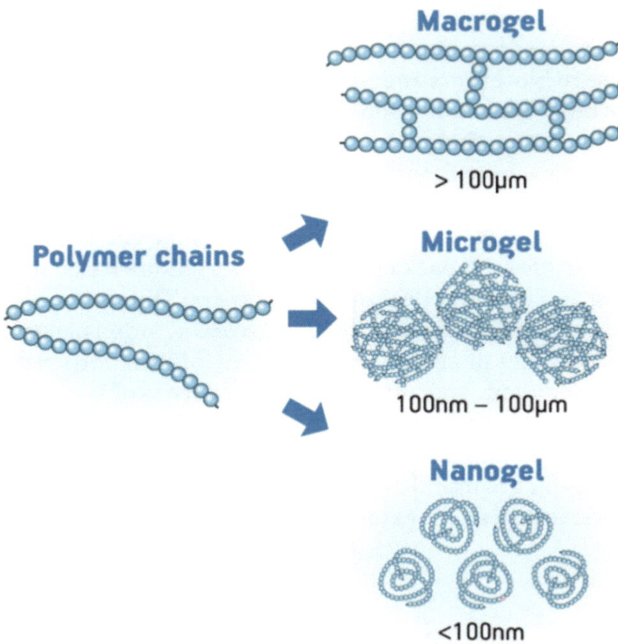

Figure 5.
Representative scheme of gels at different size levels.

lignin, their chemical interactions in producing hydrogels, and the applications and properties that make these two polymers so vital. Various applications can benefit from the incorporation of cellulose or lignin structures into hydrogels to develop unique or exceptional characteristics [15].

4.3 Agriculture

Controlling the release of macronutrients in hydrogels can be achieved by using biopolymers, which are a solution for agricultural production. To synthesize SAH, graft copolymerization of babassu mesocarp (MB) and polyacrylamide (PAM) was necessary. The characterization of these materials was done using X-ray diffraction (XRD). Differences in crystallographic characteristics between hydrogels and their precursors suggested interactions between fertilizers and macromolecular chains. FTIR revealed a band at 1121 cm^{-1} that distinguishes HMBP and HMB hydrogels and was attributed to the inclusion of potassium phosphate. The obtained materials were found to be thermally stable through thermo-gravimetric analysis (TG and DTG) [16]. The use of superabsorbent polymer hydrogels in agriculture has been presented as materials to store and retain water and nutrients and as additives to improve soil properties. The use of synthetic and natural polymer hydrogels for these purposes is considered. Although natural polymers, such as various polysaccharides, have undoubted advantages in terms of biocompatibility, biodegradability, and low cost, they are inferior to synthetic polymers in terms of water absorption and retention. The most promising in this respect are semi-synthetic polymeric superabsorbents based on natural polymers modified with additives or grafted chains of synthetic polymers, which can combine the advantages of natural and synthetic polymeric hydrogels without their disadvantages. Such semi-synthetic polymers are of great interest for agricultural applications, especially in arid regions, also because they can be used to create systems for the slow release of nutrients into the soil, which is necessary to increase crop yields using environmentally friendly technologies. One of the most pressing problems in agriculture is the shortage of water for irrigation. Superabsorbent polymers (SAPs) are widely used as containers for water and plant nutrients, especially in arid and semi-arid regions, because they significantly improve the efficiency of water use. A recent review found that applying 100 kg of SAP per hectare is the most suitable rate for increasing seed and dry matter yields while taking economic considerations into account, and it can result in a cereal seed yield increase of more than 15%. Cross-linked polymer networks that contain water-soluble building blocks are what SAPs are referred to as. Our review will categorize superabsorbent polymers or polymer hydrogels into synthetic, natural, and semi-synthetic based on their chemical structure [17]. In this review, superabsorbent polymers or polymer hydrogels have been classified according to their chemical structure as synthetic, natural, and semi-synthetic. The addition of external cross-linking molecules can lead to covalent bonding. In an aqueous environment, the dissolution of hydrophilic polymer chains can be prevented by using crosslinkers in a hydrogel preparation. Crosslinkers that are added to polymer chains can affect the physical properties of the polymer and the presence or absence of crystallinity. Cross-linking causes changes in the chemical structure of the polymer. The formation of bridges between polymer chains is possible due to the presence of at least two reactive functional groups in cross-linking molecules (**Figure 6**). Various synthetic cross-linking compounds such as dialdehydes (glyoxal, glutaraldehyde), epoxy compounds (epichlorohydrin), carbodiimides (1-ethyl-3-[3-dimethylaminopropyl] carbodiimide together with

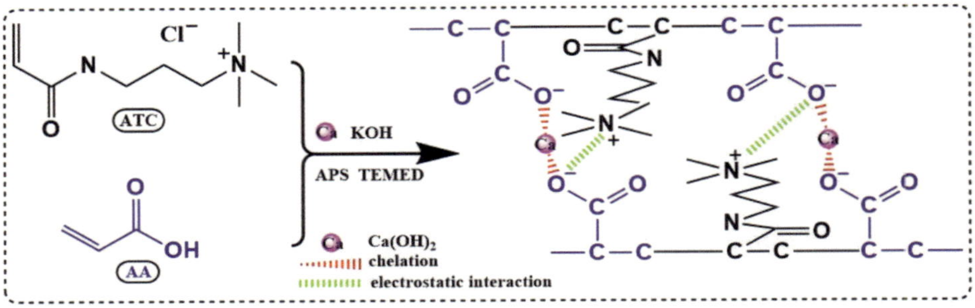

Figure 6.
A general cross-linking mechanism between nanoparticles and polymeric chains.

N-hydroxysuccinimide; EDC-NHS), and dithiols can be used for covalent cross-linking. These cross-linker molecules can be attached to polymers that have functional groups such as primary amines, alcohols, thiols, carbonyls, and carboxylic acids through covalent bonding. The Schiff base reaction allows dialdehyde crosslinkers to bind primary amines, leading to the formation of two imine bonds. NH_2-containing polysaccharides and proteins are transformed into hydrogels by the use of dialdehydes. Glyoxal dialdehyde is capable of reacting with the polysaccharide chitosan, for instance. Glutaraldehyde is regarded as one of the most efficient dialdehydes, forming stable cross-links quickly [18].

4.4 Biomedical

Advanced hydrogels can be classified as static, dynamic, multi-stage, or bioinspired. They can be used as cell-free gene expression platforms for gene therapy. Administration of nanogel-based sprays can act as an immunoassay for priming macrophage toward the M1 phenotype to prevent cancer recurrence after surgery. Nanogels can also serve as a dual biosensing and capture platform for liquid biopsies and can detect and remove circulating cancer cells from the blood of cancer patients. The ATRP technique was utilized to synthesize poly(2-hydroxyethyl methacrylate) (PHEMA)-cross-linked nanogels, which were then investigated as engineered tissue scaffolds that are biodegradable and can be cleaved by enzymes. An ABC triblock polymer (PPS-b-PDMA-b-PNIPAAM) with a reactive oxygen species (ROS)-triggered degradable response was synthesized using a combination of anionic and RAFT polymerization [19]. Collagen-based hydrogels have been the subject of a growing number of investigations, with the aim of overcoming the low mechanical properties of collagen. The biocompatibility of the collagen-alginate composite (CAC) hydrogel is widely appreciated because it gels easily under mild conditions, has low cytotoxicity, has adjustable mechanical properties, is widely available, and can be easily integrated with other biomaterials and bioactive agents. This review is designed to summarize the properties of alginate and collagen. Furthermore, the utilization of CAC hydrogel in tissue engineering and biomedical sciences is also covered [20]. A cross-linked polymer chain with a three-dimensional (3D) network structure is what hydrogels are made of. Hydrogen bonding can lead to the formation of hydrogels by the polymer with functional groups. The formation of amphiphilic grafting and block polymers can be achieved by polymers self-assembling in hydrophobic or hydrophilic solvents because of their amphiphilic affinity. The polymer chains can synthesize hydrogels by adjusting their crystallization temperature through crystallization.

Crystallization is commonly achieved through the use of the freeze-thaw and heating processes. During cross-linking, the attraction of ionic groups causes ionic interactions to occur. Proteins and polymers interact when they are added or modified using molecular medical technology, such as protein and genetic engineering. The destruction of complementary RNA strands is a result of the destruction of these small molecules, which reduces protein expression. Improving heart function and repair after myocardial infarction can be achieved through the use of miRNAs, but it requires efficient and sustained delivery to the myocardium. The use of hydrogel nanoparticles as delivery vehicles resulted in miRNA being delivered to the heart. Polymeric nanoparticles embedded in hydrogels contained miRNAs after they were loaded. MiRNA's localized and sustained release led to cardiomyocyte proliferation, which was triggered by miRNA's silencing of target proteins. The improvement of cardiac function resulted in an increase in ejection fraction, a reduction in scar size, and a double increase in capillary density in the border zone. Myocardial repair was safely and efficiently accomplished by utilizing the miRNA-loaded hydrogel nanoparticle complex (**Figure 7**) [21].

4.5 Adsorbents

The most significant environmental issues worldwide involve the presence of contaminated water and the lack of clean water. The elimination of hazardous organic/inorganic contaminants from water/sewage is crucial to protect clean water resources. Bio-based hydrogels that have luminescent properties such as three-dimensional (3D) microstructures, high porosity, surface area, and multifunctionality can be developed to meet the rising demands of wastewater treatment on an industrial scale. Advanced

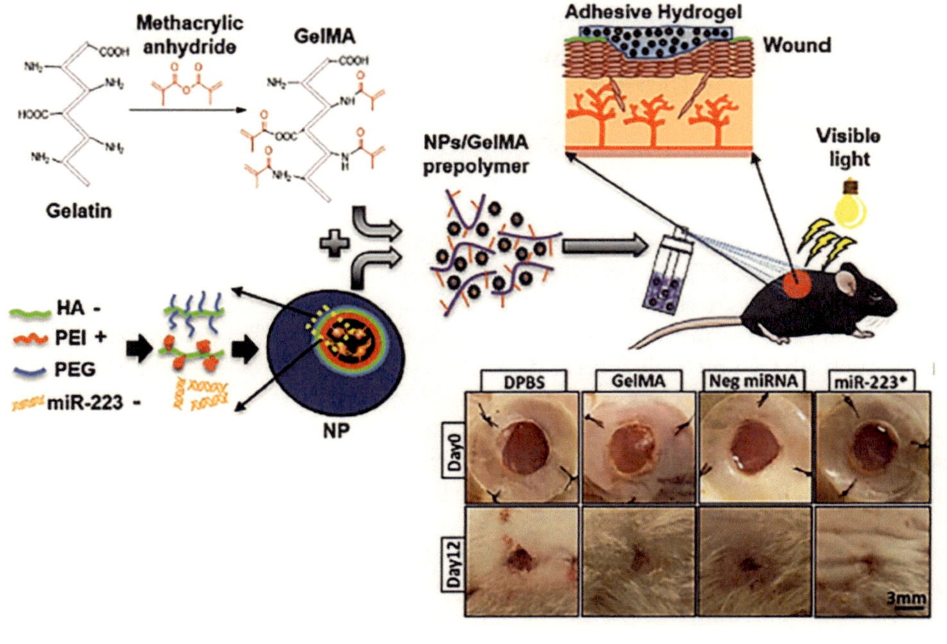

Figure 7.
Hydrogels for nucleic acid delivery. Adhesive hydrogel-loaded miRNA nanoparticles promote local immunomodulation for wound healing.

bio-based hydrogels can be developed to overcome separation problems and advance technological aspects, as described. A brief overview of each treatment technique is described as follows: (i) advanced oxidation, (ii) electrochemical method, (iii) ion exchange, (iv) biodegradation, (v) membrane filtration, (vi) flocculation/coagulation, (vii) irradiation, (viii) solar desalination, (ix) portable water filter [22]. Two cross-linked networks intertwined with balanced mechanical performance form a double-network (DN) hydrogel that has high strength and toughness. In addition to the excellent mechanical properties provided by the double-network structure, it can also have good anti-swelling and self-healing properties, which provide a good basis for the application of hydrogels in the environment. Recently, research on the environmental application of double-network hydrogels has been gradually increasing. We classify this type of hydrogel in terms of composition and cross-linking methods and discuss its adsorption mechanism for various environmental pollutants (metal ions, dyes, antibiotics). This type of hydrogel is classified in terms of composition and cross-linking methods, and its adsorption mechanism for different environmental pollutants (metal ions, dyes, and antibiotics) is discussed [23]. Polymer hydrogels have been developed in different ways to remove hazardous pollutants over the last few decades. The effectiveness of hydrogel adsorbents is determined by the gel's composition and the functions created by their polymer networks. Over the last decade, research on hydrogels that utilize the characteristic functions of polymer networks has increased. The removal of hazardous pollutants has been facilitated by the development of various adsorption functions of polymer hydrogels over the past few decades. Some adsorbents contain a local gel phase. An adsorbent that has a gel phase throughout is typically identified as a gel adsorbent. The use of polymeric hydrogels as adsorbents has several advantages. Percolation pores with the desired solvent content and pore size can be achieved through the creation of polymer networks. The network structure with a large amount of solvent makes it difficult to discuss the specific surface area of polymer hydrogels. Adsorbents can rapidly penetrate deep into the gels due to the open pores caused by the polymer networks, which enhances their sorption capacity. Gels that contain a lot of solvent have excellent diffusional permeability for a range of substances. Both experimental and theoretical studies have been conducted on the diffusivity of solutes in gels. By developing porous structures in polymeric gels, it is possible to increase the number of effective adsorption sites and decrease the resistance to mass transfer. Porous hydrogels can be synthesized using various methods, such as phase separation, emulsion, ice crystal, and foam. A solution environment that is different from the external solution can be created inside the gels due to the polymer's nature, which involves the hydrophilic-hydrophobic balance of the polymers and the presence of fixed charges. Polymer gels have been categorized as adsorbents based on their roles and functions [24]. Hydrogels are remarkably able to absorb both water and biological fluids because they are three-dimensional cross-linked stable, networks that are insoluble in water. Natural or synthetic polymers and monomers have been used to synthesize hydrogels, which have made significant progress in many different applications. The backbone of the polymer chain can be grafted with hydrophilic groups like hydroxyl (-OH), carboxylic acid (-COOH), imide (-CONH-), sulfonic acid (-SO_3H), amine (-NH_2), and amide (-$CONH_2$), and additives can also be added to create composite hydrogel. Different properties like pH, temperature, or light can stimulate polymeric composite hydrogels, which can affect swelling, mechanical properties, and self-healing, which are crucial in various fields. The creation of polymer-based composite hydrogels has been made possible by utilizing physical or chemical cross-linking techniques to enhance their physicochemical, biological, and other properties. The current focus of many researchers is on hydrogels and their applications, which include wastewater treatment and purification, medical and biomedical applications, agricultural applications, and many other industrial applications. The purpose of this review is to sum up the classification of composite hydrogels by their chemical and physical cross-linking techniques, as well as the various polymers and additives used to make them. Hydrogels have an impact on health and the environment, which is also discussed. The production and application of hydrogels have also been discussed and presented with other important issues. A hydrogel is created from a polymer substrate by using hydrophilic monomers that form double bonds during polymerization. Polymeric hydrogels can be made up of either natural or synthetic polymers, or a mixture of both. The hydrogel's mechanical, salt tolerance, and water retention properties vary depending on the monomer type. The majority of inorganic-organic DN hydrogel components consist of graphene oxide (GO) and natural minerals, among other things. Self-assembly can lead to the transformation of graphene oxide (GO) into reduced graphene oxide (rGO) hydrogels, a commonly used carbon material that contains epoxide, hydroxyl (OH) functional groups, and carboxylic acid (COOH) groups. The graphene oxide/sodium alginate DN hydrogel had an exceptional ability to absorb Mn(II) (56.49 $mg.g^{-1}$). A novel approach was taken to assemble a graphene oxide/polyacrylic acid (GO/PAA) DN gel framework, in which GO sheets self-assembled during the solution polymerization process of acrylic acid (AA) monomers; two networks were formed simultaneously (**Figure 8a**). Due to their distinctive structure and properties, hydroxyapatite and sulfoaluminate cement were utilized to construct the inorganic network of the DN hydrogels. The inorganic substance was prepared in a uniform dispersion solution after stirring, called solution A. Then appropriate amounts of monomers and other chemical reagents were mixed in water to obtain solution B. Finally, the components were completely mixed at different temperatures to form DN hydrogels. The sulfoaluminate cement/polyacrylamide

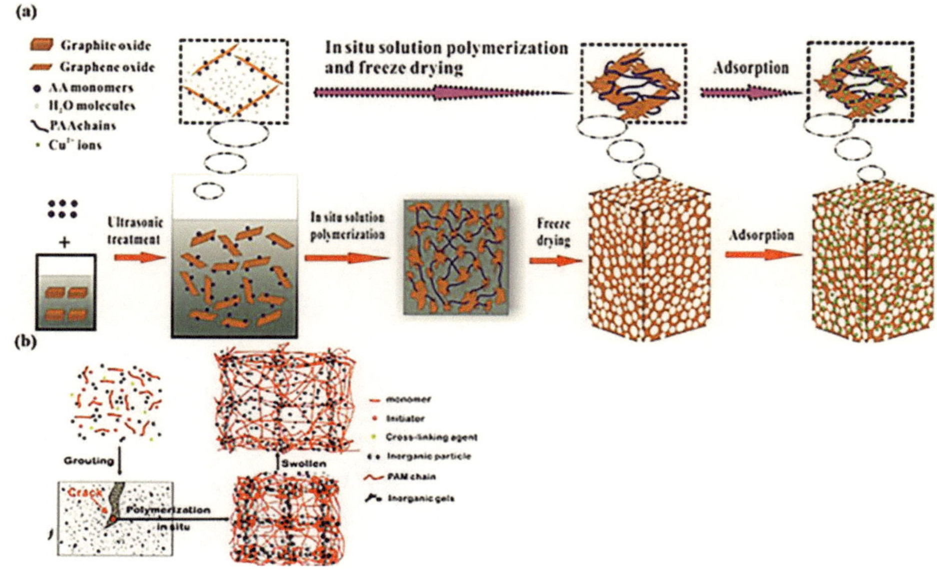

Figure 8.
Schematic of the synthesis of GO/PAA DN gel and its adsorption behavior. (a) Schematic of the synthesis of sulfoaluminate cement/polyacrylamide DN hydrogel (1) and SEM image (2) and X-ray diffraction patterns (3) (b).

DN hydrogel made through hydration and radical polymerization is extremely strong because of its distinctive network structure and special chemical composition. The network of this hydrogel is comprised of hydroxyapatite (**Figure 8b**) [25].

4.6 Ionic liquids

Biocompatibility, flexibility, and biocompatibility are all qualities that make hydrogels a natural carrier for drug delivery. Traditional PNIPAM-based hydrogel's practical application is severely hindered by its poor drug compatibility. By adding a poly (ionic liquids) chain to the binary polymer chain, we synthesized a photothermally regulated smart hydrogel. The smart hydrogel is made transparent by the temperature-induced hydrophilic-hydrophobic switchable polymer chain, which allows for intuitive visualization of drug loading and release capacity. The smart hydrogel could be used as an NIR/temperature-controlled drug carrier to achieve efficient and visualized release (40.8% drug release rate in 30 min). A dynamic wound management system was created with the smart hydrogel, which has excellent sensing performance and editable drug delivery ability, to achieve personalized wound treatment and wireless state diagnosis. Wireless monitoring and warning of joints (fingers, ankles, and elbows) could be achieved by the wireless sensor early warning system, which has guiding significance for rehabilitation treatment. Furthermore, the sensor array that uses 5 × 5 smart hydrogels could accurately detect strain, temperature, and NIR light in three-dimensional space and direction. The smart hydrogel is a new approach to personalized wound diagnosis and assisted rehabilitation. By incorporating ionic liquids into the hydrogel, it not only retains their natural benefits but also enhances its solubility and conductivity. Ultrasonic-assisted radical polymerization resulted in synthesis of a smart hydrogel that is photothermally regulated. The smart hydrogel was given editable mechanical properties, transparency, and drug-loading capacity by the addition of poly (ionic liquids) chain to the polymer network (**Figure 9**). This enabled remote, rapid, and visualized drug

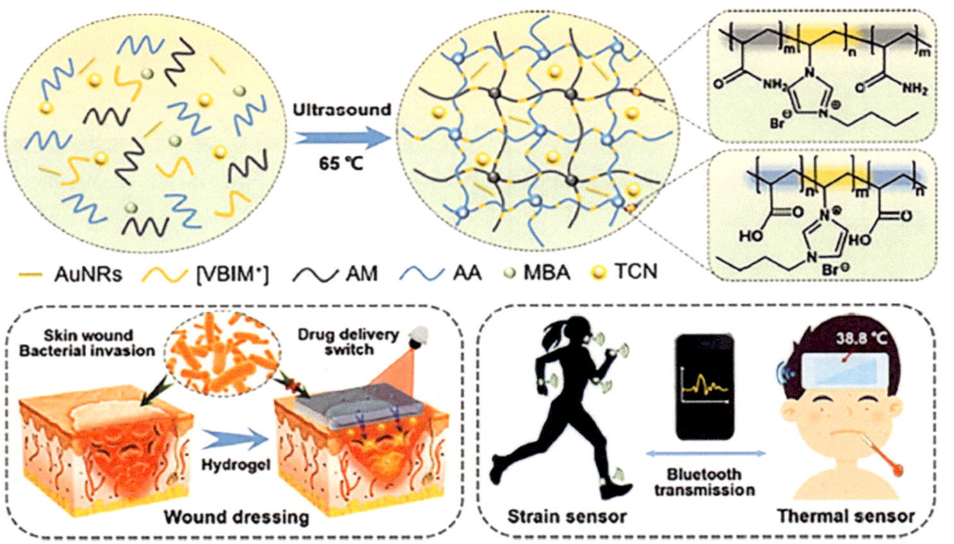

Figure 9.
Synthesis and application scheme of poly (ionic liquids) smart hydrogels.

release with a 40.8 percent release rate within 30 minutes [23]. The incorporation of ionic liquids (ILs) into hydrogel-enabled biosensing development has led to unique structures called ionic gels that have both the intrinsic attributes of hydrogels and ILs. The present review focuses on the two materials (hydrogels and ion gels) and their potential in biosensing applications in a straightforward manner. The most recent and impactful research is examined to provide a solid reference point for future development. Crosslinkers and initiators found in electrolytes play a role in the polymerization of aqueous ionic gels. Polyacrylamide-based hydrogels have been widely studied for their polymerization in the presence of KCl or NaCl-based electrolyte solutions. Polyethylene oxide and lithium salts are used to create a non-aqueous ionic gel that has recently been reported. The range of applications is limited by the low conductivity (10–5 S/cm) of lithium ions because of their mobility at room temperature. In order to improve the conductivity of ion gels, incorporating ionic liquids into polymer networks was a viable option. Dissolved salts with exceptional ionic conductivity and stability form ionic liquids (ILs) at temperatures below 100°C [26]. The report describes a hydrogel dressing that is both multifunctional and conductible, using a polymerized ionic liquid and konjac glucomannan (KGM). It was discovered that this newly developed hydrogel dressing has excellent mechanical properties and biocompatibility, as well as providing durable and efficient sterilization without releasing antibacterial factors. The current electrical stimulation was enhanced to significantly improve the migration and proliferation of fibroblast cells, as well as the treatment effect on diabetic skin wounds. A new approach to treating chronic wounds efficiently may be provided by a KGM hydrogel dressing that is polymerized and ionic liquid functionalized, along with ES. KGM, a polysaccharide that can be water soluble and flexible, can improve cellular metabolism by targeting specific sugar receptors in fibroblasts and is thought to be a promising material for building hydrogel dressing. A promising method for healing chronic wounds could involve ES and KGM-based hydrogel dressing. This purpose presents a challenge to developing hydrogel dressings based on KGM due to their poor electrical conductivity and weak antibacterial properties. The number of deaths caused by bacterial infections has risen significantly in modern healthcare, which is due to the overuse of antibiotics and the quick growth of multidrug-resistant pathogens. Numerous studies have been conducted on cationic antimicrobial polymers that have various cations, such as quaternary ammonium, phosphonium, pyridinium, and imidazole cations. PILs, which have large organic cations and small inorganic anions, are well-known for their high stability and excellent conductivity. It has been confirmed that they are promising candidates for ES treatment [27].

5. Characterization of hydrogel

The characterization of hydrogels plays a crucial role in understanding their properties and potential applications. Various analytical techniques are employed to assess the structural, mechanical, and physicochemical characteristics of hydrogels. One commonly used method is scanning electron microscopy (SEM), which provides high-resolution images of the hydrogel's surface morphology. This technique allows for the observation of the network structure, porosity, and homogeneity of the hydrogel. Additionally, Fourier-transform infrared spectroscopy (FT-IR) is used to analyze the chemical composition of hydrogels by detecting the vibrational modes of functional groups. By studying the FTIR spectra, one can identify the presence of specific chemical bonds and assess the degree of cross-linking within the hydrogel network. Light scattering techniques, such as dynamic light scattering (DLS), provide information about the size distribution and stability of hydrogel particles or aggregates in solution. Nuclear magnetic resonance (NMR) spectroscopy is another powerful tool for the characterization of hydrogels, providing insights into their molecular structure, mobility, and interactions. Furthermore, differential scanning calorimetry (DSC) can be utilized to investigate the thermal behavior of hydrogels, including the determination of phase transitions, glass transition temperature, and enthalpy changes. These techniques, including SEM, FT-IR, light scattering, NMR, and DSC, offer valuable insights into the structural, chemical, and thermal properties of hydrogels, enabling researchers to design and optimize hydrogel-based materials for various applications.

5.1 Methods to produce hydrogel

There are several methods available for the production of hydrogels, each offering unique advantages and considerations. One commonly used technique is the physical cross-linking method, where hydrogel formation relies on physical interactions, such as temperature, pH, or solvent-induced gelation. This method often involves the use of polymer chains with hydrophilic groups that can form reversible physical bonds, allowing the gel to swell and retain water. Chemical cross-linking is a method that uses cross-linking agents to create covalent bonds between polymer chains, leading to a stable network structure. This method offers excellent control over the gel's mechanical properties and stability but requires the careful selection of crosslinkers and reaction conditions. Additionally, enzymatic cross-linking has gained attention as a more environmentally friendly approach, utilizing enzymes to catalyze the formation of cross-links within the hydrogel matrix. This method offers precise control over gelation kinetics and can be used with a variety of biocompatible materials. Furthermore, recent advances in 3D printing technology have enabled the fabrication of hydrogel structures with complex geometries and spatial control. This approach combines the use of hydrogel precursors with specialized 3D printing techniques, allowing precise deposition and layer-by-layer assembly of hydrogel-based constructs. Overall, the choice of hydrogel production method depends on the desired properties, application requirements, and the compatibility of the chosen materials. The general methods to produce physical and chemical gels are described below.

5.1.1 Physical cross-linking

Physical cross-linking is a common method for the production of hydrogels, relying on physical interactions to form a three-dimensional network structure. This approach utilizes factors such as temperature, pH, or solvent-induced gelation to induce gel formation. One popular method is thermally induced gelation, where the hydrogel precursor solution undergoes a phase transition upon a change in temperature. This can be achieved by either heating or cooling the solution, causing the polymer chains to undergo conformational changes and form physical cross-links. Another approach is pH-induced gelation, where the hydrogel precursor solution contains pH-sensitive groups that can undergo protonation or deprotonation, leading to gel formation. Additionally, solvent-induced gelation involves the use of solvents that can selectively swell the hydrogel precursor, promoting the formation of physical cross-links. The advantage of physical cross-linking is that it allows for the reversible nature of the gel, as the physical bonds can be broken and reformed under certain conditions. This provides the hydrogel with the ability to respond to external stimuli and exhibit stimuli-responsive behavior. Overall, physical cross-linking is a versatile method for producing hydrogels, offering control over gelation kinetics, gel properties, and the potential for application in areas such as drug delivery, tissue engineering, and biosensors. Physically cross-linked hydrogels can be obtained using various methods, as reported in the literature. Some of these methods include:

Heating/cooling a polymer solution: the physically cross-linked gels are formed by cooling hot gelatin or carrageenan solutions, with gel formation attributed to helix formation, helix association, and the formation of junction zones. Carrageenan undergoes conformational changes upon cooling, and the addition of salt promotes the formation of stable gels. Additionally, hydrogels can be obtained through warming polymer solutions, leading to block copolymerization (**Figure 10**) [28].

Ionic interaction: The ionic polymers can be cross-linked by adding di- or trivalent counter ions, leading to gel formation. This principle is commonly applied in polyelectrolyte solutions, with examples including Na^+ alginate and various chitosan-based hydrogels. Where multivalent ions of opposite charges, such as Ca^{2+} + $2Cl^-$, are introduced to initiate cross-linking (**Figure 11**) [29].

Complex coacervation: complex coacervate gels form when a polyanion and a polycation are mixed. This is based on the principle of oppositely charged polymers binding together to create soluble or insoluble complexes, depending on solution concentration and pH (**Figure 12**). As an example, polyanionic xanthan has the ability to coacerbate with polycationic chitosan. Positively charged proteins have the ability to bind with anionic hydrocolloids below their isoelectric point to form poly-ion complex hydrogels or complex coacervates that contain poly-ion complexes [30].

5.1.2 Chemical cross-linking

Chemical cross-linking is a process that involves the formation of covalent bonds between polymer chains, resulting in the creation of a three-dimensional network structure. This network provides enhanced mechanical strength, rigidity, and stability to the material. Chemical cross-linking is achieved by introducing cross-linking agents or molecules with reactive functional groups that can form bonds with the

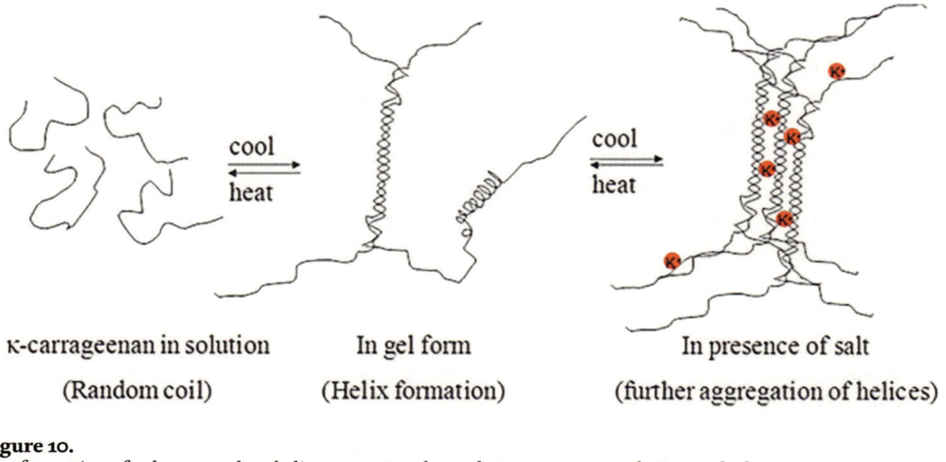

Figure 10.
The formation of gel occurs when helix aggregate when a hot carrageenan solution cools down.

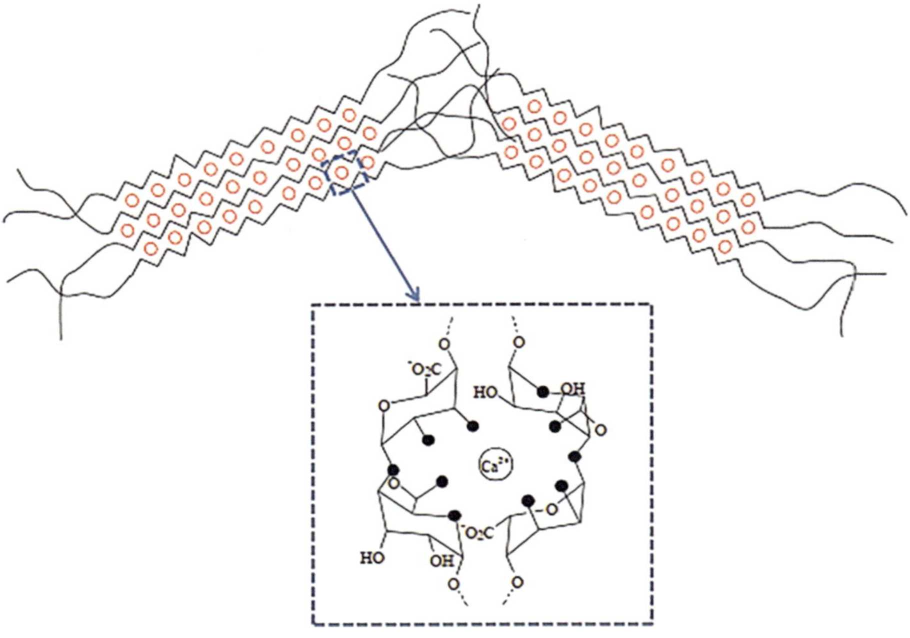

Figure 11.
Ionotropic gelation by interaction between anionic groups on alginate (COO^-) with divalent metal ions (Ca^{2+}).

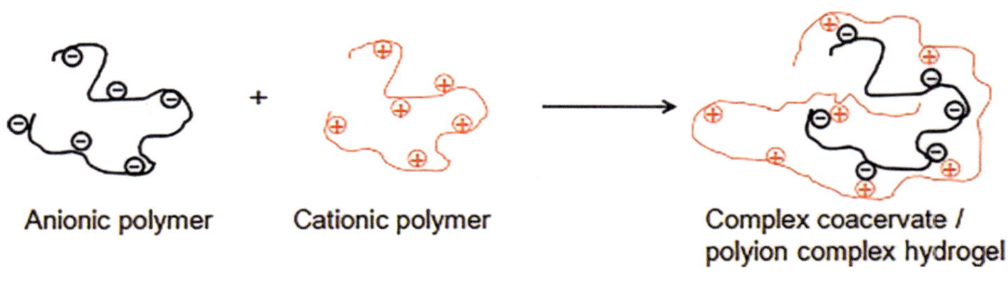

Figure 12.
Complex coacervation between a polyanion and a polycation.

polymer chains. These reactive groups can include aldehydes, isocyanates, epoxides, or vinyl groups, among others. When the cross-linking agent reacts with the polymer chains, new chemical bonds are formed, effectively connecting the chains and immobilizing them in a fixed spatial arrangement. This cross-linking process can be controlled by adjusting parameters such as cross-linker concentration, reaction time, and temperature. These agents react with functional groups present in the polymers, such as OH, COOH, or NH_2, forming covalent bonds and creating cross-links. Common crosslinkers include aldehydes like glutaraldehyde and adipic acid dihydrazide. These cross-linking reactions can occur in both natural and synthetic polymers, allowing for the modification and enhancement of their properties. Chemical cross-linking is widely used in various industries, including polymer synthesis, coatings, adhesives, and the production of biomedical materials, where it enables the development of materials with tailored properties and improved performance. As reported in the literature, crosslinkers like glutaraldehyde and epichlorohydrin are

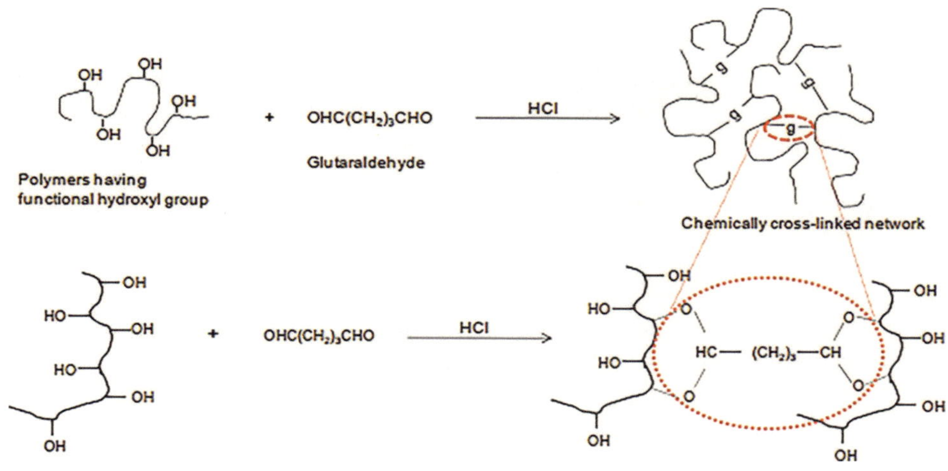

Figure 13.
Schematic illustration of using chemical cross-linker to obtain cross-linked hydrogel network.

widely used to create cross-linked hydrogel networks in both synthetic and natural polymers. This technique involves introducing additional molecules between polymer chains to form cross-linked chains (**Figure 13**). For example, glutaraldehyde can cross-link corn starch and polyvinyl alcohol to produce a hydrogel membrane with applications in artificial skin and targeted delivery of nutrients and medications [31]. Other examples include cross-linking carboxymethyl cellulose (CMC) with 1,3-diaminopropane for drug delivery, cross-linking xanthan and polyvinyl alcohol with epichlorohydrin, and cross-linking κ-carrageenan and acrylic acid using 2-acrylamido-2-methylpropanesulfonic acid for biodegradable hydrogels and drug delivery systems. Carrageenan hydrogels also show promise for industrial enzyme immobilization. Additionally, hydrogels can be synthesized from cellulose using epichlorohydrin and various methods involving NaOH/urea aqueous solutions [32, 33].

6. Conclusion

In this current chapter, the history of hydrogels, their types and properties, their formation mechanisms, and their applications in food packaging, pharmaceuticals, agriculture, biomedicine, absorbents, and ionic liquids have been discussed. In the past decades, significant progress has been made in the field of hydrogels as functional biomaterials. Today, hydrogels widely play an important role in the production of products such as contact lenses, health products, and wound dressings. However, the production of commercial hydrogels for tissue engineering and drug release is still limited. Regarding the formation mechanism of hydrogels, there are different methods for the networks of the structure. The ability to swell upon contact with an aqueous solution is a desirable property of these hydrogels. This review examines the literature on how hydrogels can be classified based on their configurations, physical and chemical characteristics, and technical feasibility of utilization. Our review of nanoparticle-hydrogel composites has been extensive, as they are a state-of-the-art and versatile class of materials that can be used in a variety of applications. Many scaffolds and drug release tools based on hydrogels have been designed, studied, and, in some cases, even patented, but few of them have reached the market. Further development for the production of commercial hydrogels is expected in these two fields, but high production costs are the main reason for the limitation of their further commercialization. Recent advances in polymer science and technology have led to the development of various types of hydrogels. But despite all the beneficial properties of hydrogels, there are still many challenges to overcome for clinical transfer. Future research in this field will be focused on the transfer of promising preclinical studies and bioprocesses to change life with the ability to increase the quality of life and healthy aging on a global scale. The classification of synthetic approaches and applications of nanoparticle-hydrogel composites presented in this review is anticipated to enhance the reader's understanding of the system and facilitate the design of innovative combinations for novel applications.

Acknowledgements

The authors appreciate the support from the University of Mazandaran. Also, the generous contributions of all our collaborators over the years are appreciated.

Conflict of interest

The authors declare no conflict of interest.

Author details

Behrooz Maleki[1*], Pouya Ghamari Kargar[1], Samaneh Sedigh Ashrafi[1] and Milad Ghani[2]

1 Faculty of Chemistry, Department of Organic Chemistry, University of Mazandaran, Babolsar, Iran

2 Faculty of Chemistry, Department of Analytical Chemistry, University of Mazandaran, Babolsar, Iran

*Address all correspondence to: b.maleki@umz.ac.ir

References

[1] Alaei J, Boroojerdi SH, Rabiei Z. Application of hydrogels in drying operation. Petroleum and Coal. 2005;**47**(3):32-37

[2] Sapuła P, Bialik-Wąs K, Malarz K. Are natural compounds a promising alternative to synthetic cross-linking agents in the preparation of hydrogels? Pharmaceutics [Internet]. 2023;**15**(1):253. Available from: https://www.mdpi.com/1999-4923/15/1/253

[3] Correa S, Grosskopf AK, Lopez Hernandez H, Chan D, Yu AC, Stapleton LM, et al. Translational applications of hydrogels. Chemical Reviews [Internet]. 2021;**121**(18):11385-11457. DOI: 10.1021/acs.chemrev.0c01177

[4] Thakur S, Thakur VK, Arotiba OA. History, Classification, Properties and Application of Hydrogels: An Overview. Singapore: Springer; 2018. pp. 29-50. DOI: 10.1007/978-981-10-6077-9_2

[5] Lee SC, Kwon IK, Park K. Hydrogels for delivery of bioactive agents: A historical perspective. Advanced Drug Delivery Reviews. 2013;**65**(1):17-20. Available from: https://linkinghub.elsevier.com/retrieve/pii/S0169409X12002451

[6] Buwalda SJ, Boere KWM, Dijkstra PJ, Feijen J, Vermonden T, Hennink WE. Hydrogels in a historical perspective: From simple networks to smart materials. Journal of Controlled Release [Internet]. 2014;**190**:254-273. Available from: https://linkinghub.elsevier.com/retrieve/pii/S0168365914002181

[7] Bustamante-Torres M, Romero-Fierro D, Arcentales-Vera B, Palomino K, Magaña H, Bucio E. Hydrogels classification according to the physical or chemical interactions and as stimuli-sensitive materials. Gels [Internet]. 2021;7(4):182. Available from: https://www.mdpi.com/2310-2861/7/4/182

[8] Aswathy SH, Narendrakumar U, Manjubala I. Commercial hydrogels for biomedical applications. Heliyon [Internet]. 2020;**6**(4):e03719. Available from: https://linkinghub.elsevier.com/retrieve/pii/S2405844020305648

[9] Kasai RD, Radhika D, Archana S, Shanavaz H, Koutavarapu R, Lee DY, et al. A review on hydrogels classification and recent developments in biomedical applications. International Journal of Polymeric Materials and Polymeric Biomaterials [Internet]. 2023;**72**(13):1059-1069. DOI: 10.1080/00914037.2022.2075872

[10] Brinchi L, Cotana F, Fortunati E, Kenny JM. Production of nanocrystalline cellulose from lignocellulosic biomass: Technology and applications. Carbohydrate Polymers. 2013;**94**(1):154-169

[11] Yang J, Shen M, Luo Y, Wu T, Chen X, Wang Y, et al. Advanced applications of chitosan-based hydrogels: From biosensors to intelligent food packaging system. Trends in Food Science & Technology [Internet]. 2021;**110**:822-832. Available from: https://linkinghub.elsevier.com/retrieve/pii/S0924224421001394

[12] Lu P, Yang Y, Liu R, Liu X, Ma J, Wu M, et al. Preparation of sugarcane bagasse nanocellulose hydrogel as a colourimetric freshness indicator for intelligent food packaging. Carbohydrate Polymers [Internet]. 2020;**249**:116831. Available from: https://

linkinghub.elsevier.com/retrieve/pii/S0144861720310043

[13] Leyva-Jiménez FJ, Oliver-Simancas R, Castangia I, Rodríguez-García AM, Alañón ME. Comprehensive review of natural based hydrogels as an upcoming trend for food packing. Food Hydrocolloids [Internet]. 2023;**135**:108124. Available from: https://linkinghub.elsevier.com/retrieve/pii/S0268005X22006440

[14] CSA DL, Balogh TS, JPRO V, GHC V, ABA L, Camacho-Cruz L, et al. An updated review of macro, micro, and nanostructured hydrogels for biomedical and pharmaceutical applications. Pharmaceutics [Internet]. 2020;**12**(10):970. Available from: https://www.mdpi.com/1999-4923/12/10/970

[15] Barzegar S, Aryaie Monfared MH, Hubbe MA. Cellulose and lignin as propitious candidates for preparation of hydrogels for pharmaceutical applications. Materials Today Communications [Internet]. 2022;**33**:104617. Available from: https://linkinghub.elsevier.com/retrieve/pii/S2352492822014581

[16] Silva AO, Cunha RS, Hotza D, Machado RAF. Chitosan as a matrix of nanocomposites: A review on nanostructures, processes, properties, and applications. Carbohydrate Polymers [Internet]. 2021;**272**:118472. Available from: https://linkinghub.elsevier.com/retrieve/pii/S0144861721008596

[17] Krasnopeeva EL, Panova GG, Yakimansky AV. Agricultural applications of superabsorbent polymer hydrogels. International Journal of Molecular Sciences [Internet]. 2022;**23**(23):15134. Available from: https://www.mdpi.com/1422-0067/23/23/15134

[18] Klein M, Poverenov E. Natural biopolymer-based hydrogels for use in food and agriculture. Journal of the Science of Food and Agriculture [Internet]. 2020;**100**(6):2337-2347. DOI: 10.1002/jsfa.10274

[19] Ayoubi-Joshaghani MH, Seidi K, Azizi M, Jaymand M, Javaheri T, Jahanban-Esfahlan R, et al. Potential applications of advanced nano/hydrogels in biomedicine: Static, dynamic, multi-stage, and bioinspired. Advanced Functional Materials [Internet]. 2020;**30**(45):2004098. DOI: 10.1002/adfm.202004098

[20] Hu T, Lo ACY. Collagen–Alginate composite hydrogel: Application in tissue engineering and biomedical sciences. Polymers (Basel) [Internet]. 2021;13(11):1852. Available from: https://www.mdpi.com/2073-4360/13/11/1852

[21] Ho TC, Chang CC, Chan HP, Chung TW, Shu CW, Chuang KP, et al. Hydrogels: Properties and applications in biomedicine. Molecules [Internet]. 2022;**27**(9):2902. Available from: https://www.mdpi.com/1420-3049/27/9/2902

[22] Godiya CB, Martins Ruotolo LA, Cai W. Functional biobased hydrogels for the removal of aqueous hazardous pollutants: Current status, challenges, and future perspectives. Journal of Materials Chemistry A [Internet]. 2020;**8**(41):21585-21612. Available from: http://xlink.rsc.org/?DOI=D0TA07028A

[23] Qu X, Liu J, Wang S, Shao J, Wang Q, Wang W, et al. Photothermal regulated multi-perceptive poly(ionic liquids) hydrogel sensor for bioelectronics. Chemical Engineering Journal [Internet]. 2023;**453**:139785. Available from: https://linkinghub.elsevier.com/retrieve/pii/S1385894722052640

[24] Seida Y, Tokuyama H. Hydrogel adsorbents for the removal of hazardous pollutants—Requirements and available

functions as adsorbent. Gels [Internet]. 2022;**8**(4):220. Available from: https://www.mdpi.com/2310-2861/8/4/220

[25] El Sayed MM. Production of polymer hydrogel composites and their applications. Journal of Polymers and the Environment [Internet]. 2023;**31**(7):2855-2879. DOI: 10.1007/s10924-023-02796-z

[26] Ghorbanizamani F, Moulahoum H, Guler Celik E, Timur S. Ionic liquids enhancement of hydrogels and impact on biosensing applications. Journal of Molecular Liquids [Internet]. 2022;**357**:119075. Available from: https://linkinghub.elsevier.com/retrieve/pii/S0167732222006134

[27] Liu P, Jin K, Wong W, Wang Y, Liang T, He M, et al. Ionic liquid functionalized non-releasing antibacterial hydrogel dressing coupled with electrical stimulation for the promotion of diabetic wound healing. Chemical Engineering Journal [Internet]. 2021;**415**:129025. Available from: https://linkinghub.elsevier.com/retrieve/pii/S1385894721006185

[28] Funami T, Hiroe M, Noda S, Asai I, Ikeda S, Nishinari K. Influence of molecular structure imaged with atomic force microscopy on the rheological behavior of carrageenan aqueous systems in the presence or absence of cations. Food Hydrocolloids [Internet]. 2007;**21**(4):617-629. Available from: https://linkinghub.elsevier.com/retrieve/pii/S0268005X06001500

[29] Bajpai AK, Shukla SK, Bhanu S, Kankane S. Responsive polymers in controlled drug delivery. Progress in Polymer Science [Internet]. 2008;**33**(11):1088-1118. Available from: https://linkinghub.elsevier.com/retrieve/pii/S0079670008000609

[30] Magnin D. Physicochemical and structural characterization of a polyionic matrix of interest in biotechnology, in the pharmaceutical and biomedical fields. Carbohydrate Polymers [Internet]. 2004;**55**(4):437-453. Available from: https://linkinghub.elsevier.com/retrieve/pii/S0144861703003199

[31] Hennink WE, van Nostrum CF. Novel crosslinking methods to design hydrogels. Advanced Drug Delivery Reviews [Internet]. 2012;**64**:223-236. Available from: https://linkinghub.elsevier.com/retrieve/pii/S0169409X12002694

[32] Chang C, Zhang L, Zhou J, Zhang L, Kennedy JF. Structure and properties of hydrogels prepared from cellulose in NaOH/urea aqueous solutions. Carbohydrate Polymers [Internet]. 2010;**82**(1):122-127. Available from: https://linkinghub.elsevier.com/retrieve/pii/S0144861710003000

[33] Chang C, Zhang L. Cellulose-based hydrogels: Present status and application prospects. Carbohydrate Polymers [Internet]. 2011;**84**(1):40-53. Available from: https://linkinghub.elsevier.com/retrieve/pii/S0144861710010003

Chapter 9

Recent Advances and Challenges of Ionic Liquids in Tribology

Hong Guo, Junru Pang and Patricia Iglesias

Abstract

Ionic liquids with unique physical-chemical properties have shown their potential use as lubricants with promising lubricity. The aim of this chapter is to summarize the category of ionic liquids that have been studied in tribology. The advances of ionic liquids in macro- and nano-scale tribology regarding the lubrication behaviors and mechanisms are discussed. Their physico-chemical properties such as viscosity, thermal stability, conductivity, biodegradability, toxicity are reviewed. The effect of humidity and interface forces on adhesion, frictional behavior, and lubricity of ionic liquids, particularly in nanoscale, are discussed. The challenges and possible future research directions are concluded at the end of this chapter.

Keywords: ionic liquid, lubrication, friction, tribology, nanotribology

1. Introduction

Ionic Liquids (ILs) have attracted interest in the chemical community over the past few decades due to their extraordinary physicochemical properties [1–3]. The possibility of tailoring their properties to fit the needs of a particular application by selecting the appropriate cations and anions makes them very attractive to replace volatile industrial solvents [4–7]. It was not until 2001, when Ye et al. [8] published the first research article about the use of ILs in lubrication. In particular, two imidazolium tetrafluoroborate ILs were used as neat lubricants in different frictional pairs and compared with two traditional lubricants. Both ILs showed friction and wear reduction under the conditions studied compared to the traditional lubricants, owing its better lubricating ability to the formation of an effective boundary film between the surfaces in contact. After the publication of these groundbreaking results, interest in these ionic fluids started to grow in the field of tribology. Early studies focused on aprotic ionic liquids (APILs) with halogen-containing anions [9–12]. However, it was soon recognized that the complex route of synthesizing APILs and their practicality were limited by high costs. This challenge was later addressed by incorporating ILs as additives to lubricants. In 2003, a pioneering study was published [13], utilizing imidazolium tetrafluoroborate ILs as additives to lubricants. Since then, numerous studies have been published using ILs as additives to different base lubricants. Another challenge encountered in the commercialization of ILs was the presence of halogens in the structure of early-used ILs [14–16]. When these ILs come into contact with water, they produce hydrohalic acid, which is highly toxic to the environment and can cause corrosion on metal surfaces. With increasingly more environmental regulations and growing environmental awareness, recent studies avoid halogen elements in the IL's structure. Moreover, the emerging trend in this field involves the utilization of protic ionic liquids (PILs), which are easier to synthesize through a proton transfer from an acid to a base, which results in lower costs.

Compared to traditional lubricants, ILs have several advantages. However, it needs careful investigation to understand how their properties influence their tribological performance as lubricants or lubricant additives. Early studies focused on identifying ILs with suitable viscosity and miscibility with traditional lubricants to improve their performance [17]. As research progressed, the focus shifted toward understanding how specific physico-chemical properties relate to friction reduction. Viscosity emerged as a crucial factor, with moderate viscosity offering a balance between load-carrying capacity and ease of flow. Additionally, the polarity and ionicity of ILs were found to influence their ability to form thin films on metal surfaces, reducing friction through adsorption and boundary lubrication mechanisms [18, 19]. Recent studies dive deeper into how tailoring specific properties unlocks even better tribological performance. Some research has explored how functional groups on the IL cations or anions interact with surfaces. These interactions lead to the formation of tribochemical films that minimize wear [20]. Similarly, the choice of counterion can influence the IL's interaction with water, impacting friction and preventing corrosion [21]. Beyond traditional properties, researchers are investigating novel approaches to optimize ILs for lubrication. The addition of nanoparticles or organic modifiers can further enhance film formation, reduce friction, and improve load-carrying capacity [22]. Furthermore, incorporating self-healing functionalities allows ILs to repair damaged films, extending their lifetime and enhancing their overall performance.

As research on ILs for lubrication advances, understanding their behavior at the nanoscale has become increasingly crucial. This realm of nanoscale tribology offers unique insights into the fundamental mechanisms influencing lubrication performance and paves the way for optimized IL design. With advancements in techniques like in-situ tribometers [23] and molecular dynamics simulations [24], researchers further investigated the dynamic phenomena governing friction reduction. The formation of tribofilms is detected on metal surfaces due to IL adsorption and tribochemical reactions from many studies. Understanding how these films evolve under pressure and shear, influenced by factors like chain length and functional groups, became paramount.

In this chapter, several physico-chemical properties of ionic liquids including viscosity, thermal stability, conductivity, biodegradability, and toxicity are summarized. The recent development of ionic liquid in macroscale and nanoscale tribology is also reviewed. We also offer a glimpse into the future of ionic liquid's development.

2. Physico-chemical properties of ionic liquids

Physico-chemical properties of ionic liquids are commonly considered as the vital parameters for understanding and optimizing their performance as lubricants. These properties influence various lubrication mechanisms, affecting friction, wear, and film formation, ultimately determining the effectiveness of ILs in tribological applications. Meanwhile, the evaluation of toxicity and biodegradability plays a crucial role in determining if ILs are eco-friendly and suitable as a substitute for commercial lubricants currently in use (**Figure 1**).

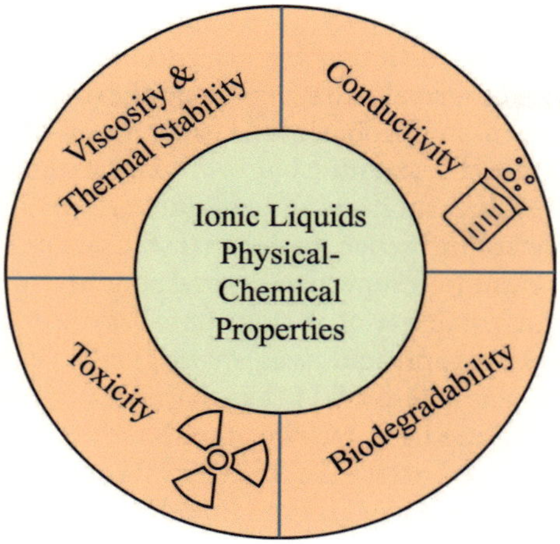

Figure 1.
Graphic abstract for physical-chemical properties of ionic liquids.

2.1 Viscosity and thermal stability

One of the vital challenges in developing high performance lubricants is maintaining optimal viscosity under harsh operating conditions [25]. This is particularly crucial for designing innovative lubricants like ILs. While ILs offer several advantages, it is still a critical research area to control their viscosity for efficient lubrication.

A good understanding of how the structure of ILs, such as the size, charge density, and molecular shape of both cations and anions, influences their viscosity has been established. Increasing the chain length of the alkyl substituents on the cations will increase viscosity [26]. Meanwhile, ILs with bulky and complex anions tend to have higher viscosity compared to the ones with smaller and simpler anions. Researchers could develop ILs with optimal viscosities suitable for target lubrication applications by choosing and tailoring the cation and anion combinations [27].

The negligible vapor pressure is another advantage of ILs, which could eliminate concerns about lubricant evaporation under harsh operating conditions [28]. This feature makes ILs more environmentally friendly and reduces potentially the need for frequent lubricant refills. However, thermal degradation of lubricants under extreme heat conditions is considered as one of the possible reasons for viscosity variation and lubrication failure [29]. Thus, the thermal stability of ILs is a critical parameter to estimate and evaluate the tribological behavior when used as lubricants or lubricant additives. Blanco et al. [30] studied the thermal stability of three bis(trifluoromethylsulfonyl) amide ([NTf_2]) based ILs, namely [$C_{12}mim$][NTf_2], [N_{4441}][NTf_2], [N_{8881}][NTf_2]. They discovered that the anion and cation nature have a great influence on IL heat dislocation. Lengthening the chain reduces intramolecular electrostatic interaction, thereby lowering thermal stability. Specifically, the imidazolium IL, represented by [$C_{12}mim$] [NTf_2], exhibited the highest thermal stability among the three ILs. Moreover, the imidazolium cation had an activation energy of 125 kJ/mol, compared to 121 kJ/mol and 108 kJ/mol for [N_{4441}][NTf_2] and [N_{8881}][NTf_2], respectively, indicating a lower thermal degradation for IL with imidazolium cation. Liaw et al. [31] investigated the flashpoint of two imidazolium ILs, ([C_6mim][Cl]) and ([C_2mim][NTf_2]), at various temperatures. Increasing the temperature reduced the flash temperature of [C_6mim][Cl]. Additionally, the flash temperature of [C_2mim][NTf_2] substantially decreased at 200°C.

Parajo et al. [32] proposed four novel ILs, incorporating two different cations (pyridinium and choline) and three different anions (tosylate, acetate, and dihydrogen phosphate). They analyzed thermal behaviors using thermogravimetric techniques and studied phase transitions and isobaric heat capacities with differential scanning calorimetry. The ILs exhibited crystalline behavior with melting temperatures ranging from 88°C to 123°C. Heat capacities increased linearly with temperature, with the anion showing no significant influence on thermal stability or heat capacity. Notably, ILs with tosylate anions exhibited superior thermal stability and heat capacity.

In Guo [18], the thermal stability of various families of protic ionic liquids (PILs) and their corresponding bases and acids was investigated. The study revealed that the thermal stability of the hexanoate-based PILs is significantly influenced by thermal stability of their corresponding base. On the other hand, for p-toluenesulfonate-based and dodecylbenzenesulfonate-based PILs, the thermal stability of the corresponding acid appeared to have a greater impact on the final PILs' thermal stability.

2.2 Conductivity

Compared with conventional lubricants, ILs have unique electrical conductivity. However, it also leads to concerns about potential electrical shorts and interference in machinery with electrical components. At the same time, the potential benefits of conductivity gained attention, such as influencing the formation of tribofilms and facilitating electrostatic interaction with surfaces.

Recent research has discovered several factors which influence conductivity, including:

- Ionic structure: the size and charge density of the cations and anions play a significant role. Smaller and less complex ions normally lead to a higher conductivity [33].

- Temperature: conductivity often increases with temperature because mobility of charged ions is increased with temperature [34].

- Concentration: the concentration of ions in the IL directly affects conductivity. Higher concentrations generally lead to a higher conductivity [35].

Researchers also proposed several strategies to ensure ILs with the desired conductivity for improving tribological performance. In Guo et al. [36], ILs with specific cation and anion combinations are investigated, which offer both desirable lubrication properties and moderate conductivity to reduce the risk of electrical issues. Mixing ILs with non-conductive base oils is also another efficient way to control the dilution of conductivity and keep the beneficial properties of ILs [37]. Recently, to modify surfaces with materials like graphene [38] or carbon nanotubes [39] are proposed to be an alternative way to create conductive pathways and minimize the need for highly conductive ILs.

In addition, ILs with high thermal conductivity would help to dissipate heat during sliding movement. Similar to electrical conductivity, several factors have been discussed to potentially influence the thermal conductivity of ILs:

- Ionic structure: the size and complexity of both cations and anions affect their thermal conductivity. Smaller and simpler ions are considered to facilitate easier heat transfer, which leads to a higher thermal conductivity [40].

- Viscosity: there is a strong correlation between viscosity and thermal conductivity. Higher viscosity often implies a lower thermal conductivity since the mobility of ions of the IL is decreased [41].

- Temperature dependence: there is no clear relationship between thermal conductivity of ILs and the operating temperatures. Some ILs show a slight increase in thermal conductivity with increasing temperature, while others reveal a decrease [42]. The reason of the opposite temperature dependence is not clear yet.

Recently, researchers introduced specific functional groups into the structure of ILs, which potentially influence their thermal conductivity, offering a targeted approach for optimization [43].

2.3 Biodegradability

Biodegradability, the breakdown of complex organic molecules into simpler compounds by microorganisms, is crucial for minimizing the environmental impact of large-scale ionic liquid usage [44]. Microbial populations with diverse catabolic capabilities govern this process, similar to how they influence toxicity.

Several standard tests (e.g., OECD 301, ASTM D5988, ISO 14593) are applied to assess IL biodegradability in water and soil. The OECD 301 A-F method classifies biodegradability as "Readily biodegradable" (≥60% breakdown by activated sludge), "Inherently biodegradable" (>20% but <60%), or "Not readily biodegradable" (<20%) within a 10-day window of a 28-day test [45]. While convenient, this method lacks direct chemical analysis or immediate toxicity measures.

Researchers have demonstrated the low or non-biodegradability of ILs is determined by side chain size, central ring structure and the presence of functional groups [46]. Chain length plays a significant role in the biodegradability of ILs: those with alkyl chains exceeding 6–8 carbon atoms exhibit some degree of breakdown [47]. However, head group composition also impacts biodegradability. While imidazolium based ILs tend to resist degradation, pyridinium ILs show promising potential, especially when containing ester substituents in specific positions [48]. Similarly, phosphonium-based ILs demonstrate a good level of biodegradability [44].

In Viesca et al. [49], researchers assess the biodegradability of six protic ionic liquids using chemical oxygen demand (COD) and biochemical oxygen demand (BOD5). Six PILs include three hexanoate-based PILs: 2-hydroxyethylammonium 2-ethylhexanoate (PIL-1), 2-hydroxymethylammonium 2-ethylhexanoate (PIL-2), and 2-hydroxydimethylammonium 2-ethylhexanoate (PIL-3), and the three sulfonate-based PILs: 2-hydroxyethylammonium 4-dodecylbenzenesulfonate (PIL-4), 2-hydroxymethylammonium 4-dodecylbenzenesulfonate (PIL-5), and 2-hydroxydimethylammonium 4-dodecylbenzenesulfonate (PIL-6). Generally, sulfonate-based PILs show lower biodegradability, which is related to the presence of the benzenesulfonate aromatic group in its anionic part [50]. Compared with reference lubricant additive (ZDDP), all PILs exhibit low levels of biodegradation, which reveals the great potential of PILs used as green lubricant additives.

2.4 Toxicity

The toxicity of ILs has also gained the attention of the tribology field. Recently, there are numerous studies to discuss the toxicity and environmental impact of ILs on water and soil [51, 52]. Several methodologies exist for evaluating ionic liquid's toxicity, but the OECD proposals are widely considered the most reliable and recommended. These methods utilize the half maximal effective concentration (EC50) as the key toxicity index, providing valuable insights into potential human health and environmental impacts [53]. Notably, the Registration, Evaluation, Authorization, and Restriction of Chemicals (REACH) regulatory framework leverages these very methods for risk assessment and classification of ILs. Although the mechanism of IL toxicity is not fully clear, several publications have proposed that the structure of the cation of ILs has more influence on the toxicity than the structure of the anion [54]. Studies reveal a direct link between longer alkyl chains in ILs and increased toxicity and degradation rates. This connection stems from the chain's tendency to form non-aqueous phases or micelles within physiological environments. These structures can disrupt biological processes, leading to both toxic effects and enhanced accessibility for degrading organisms.

Compared to traditional APILs, PILs exhibit lower toxicity [55]. Consequently, researchers have actively synthesized PILs with shorter alkyl side chains [56]. These newly developed PILs demonstrate reduced toxicity and, significantly, stimulate greater microbial activity in soil. This enhanced microbial activity suggests their potential biodegradability, offering an enticing advantage over their APIL counterparts.

3. The macroscale tribology of ILs

Figure 2 illustrates the advancements in ILs' research in the field of tribology. The following section of this chapter will provide a brief review of each of the stages illustrated in the figure.

3.1 Aprotic ionic liquids

ILs can be divided into two main categories: APILs and PILs. APILs are synthesized following a complex and with costly two-steps route [57]: (1) formation of the cation via quaternization of amines or phosphines with alkylation agents; and

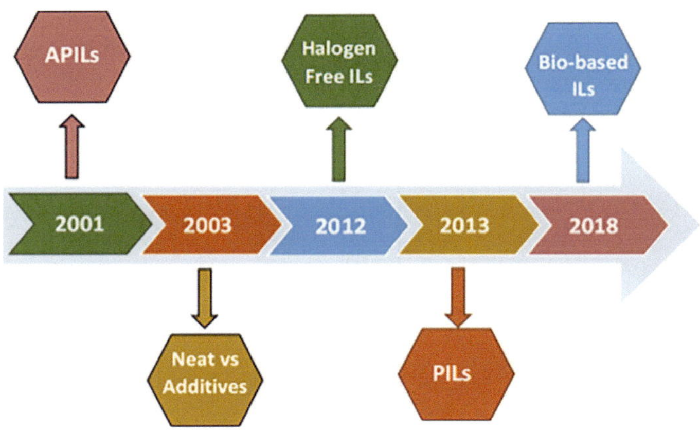

Figure 2.
Advancements in ILs' research in the field of tribology.

(2) anion exchange. In the early works within the field of tribology, the majority of ILs utilized were APILs. Organic cations such as imidazolium [12, 15, 58], pyridinium [59], ammonium [11, 15, 60, 61], and phosphonium [62, 63], were frequently paired with weakly coordinating anions such as tetrafluoroborate [64–66], hexafluorophosphate [67], bis(fluoroalkylsulfonyl)amides [65, 68], and sulfonates [69, 70] for applications in steel-steel [71, 72], aluminum-steel [10, 73, 74], steel-copper [75, 76], titanium-steel [61, 77] and others [60, 78]. Friction reductions often resulted from the formation of a physically absorbed layer between the anionic moieties and the metal surfaces [5, 79]. Under appropriate conditions of temperature and pressure, a tribofilm was also created on the surface of the metals in contact, as a result of the tribochemical reactions between the ILs and the lubricated surfaces [5, 79].

3.2 Ionic liquids lubricant additives

Although the preliminary results of using ILs as neat lubricants were promising [8, 9, 65], with significant friction and wear reductions compared to commercially available lubricants at the time, a major limitation of their use was their high cost derived from the complex synthesis route. In 2003, a study was published using ILs for the first time as additives to lubricants [13]. Subsequent works continued studying ILs as additives and found that, in most cases, adding small amounts of ILs to base oils not only would not increase the cost of the lubricant mixture [58], but it could also lead to important friction and wear reductions. Concentrations ranging from 0.5 to 5 wt.% [73, 80] were considered optimum for friction and wear reductions, but specific amount is influenced by the base oil used, the load, the speed, the materials in contact and of course, the chemical structure of the IL. Jimenez et al. [58] found that the use of ILs contacting short alkyl chain lengths and more reactive anions results in high friction and wear values for aluminum-steel contacts due to tribocorrosion processes between the ILs and the surfaces in contact. However, when these ILs are used as 1 wt.% additives to a mineral oil, the absence of tribocorrosion reactions leads to low friction values throughout the entire tests. In general, longer alkyl chain lengths and less reactive anions [58] are desired when the IL is going to be used as a neat lubricant, and shorter alkyl chain lengths and more reactive anions are desired when they are going to be used as additives.

The amount of IL used as additive is also often limited by the solubility of the ILs in the base oil [62], particularly in non-polar hydrocarbons oils. Jun Qu's group extensively studied [62, 70, 81] the solubility of ionic liquids (ILs) in mineral and synthetic oils. They found that phosphonium-based ILs with long alkyl chain lengths and quaternary structures in their cations and anions are promising candidates to be used as additives in non-polar hydrocarbon oils. The excellent miscibility observed in this family of phosphonium-based ILs is attributed to the presence of long alkyl chains and the ability to form an H−O bond between the cation and the anion. In a more systematic study conducted by this group, exploring cations with varying alkyl chain lengths, it was determined that a six-carbon chain is the critical minimum length for achieving high miscibility in non-polar base oils [82].

3.3 Halogen-free ionic liquids

In the early stages, ILs containing halogen atoms in their anion were widely used. These highly reactive elements have the potential to react with metals, forming tribolayers on the surfaces to protect them again friction and wear [83, 84].

However, in certain scenarios, an excess of these reactions can, counterintuitively, accelerate the wear process rather than providing protection to the surfaces, ultimately promoting corrosion. It is also well known that ILs with trifluoroborate (BF4-), hexafluorophosphate (BF6-) and other halogen-containing anions can undergo hydrolysis in the presence of moisture, resulting in the formation of toxic byproducts [85]. Addressing both of these challenges simultaneously, and with environmental considerations in mind, interest in halogen-free ionic liquids within the lubrication community began in the second decade of the twenty-first century. In [86], six halogen-, sulfur-, and phosphorus-free ionic liquids (ILs) demonstrated significant reductions in friction and wear compared to a polyalphaolefin oil under a boundary lubrication regime. This was attributed to the formation of tribolayers rich in iron oxide, iron carboxylates, and complex compounds. Additionally, these six ILs exhibited non-corrosive characteristics and showed no signs of corrosion under corrosion strip tests, making them potential candidates for effective lubricants. In [87], a tribo-pair that is challenging to lubricate, such as aluminum-steel, was effectively lubricated with a halogen-free ionic liquid, resulting in a low friction coefficient and almost negligible wear. In general, multiple studies have demonstrated that halogen-free ionic liquids can be used as effective lubricants and lubricant additives, avoiding corrosion problems and the toxicity associated with halogen-containing counterparts [87–89].

3.4 Protic ionic liquids

A few papers were published before [90] but it was not until 2013 that significant advances were made in the area of PILs for use as lubricants and lubricant additives. PILs are a subgroup of ILs that can easily been synthesized by a combination of a Bronsted acid with a Bronsted base [90]. Their cost-efficient synthesis route, in combination with the potential to be designed without halogen or other undesirable elements like phosphorus or sulfur, makes them ideal candidates to replace the conventional and costly APILs. PILs possess unique properties that distinguish them from APILs due to the presence of available protons in their molecules, resulting in the potential to form hydrogen bonding [90, 91]. Among other properties, the potential to form hydrogen bonds may improve the solubility of the PIL in the base oil [92]. In PILs, different strength acids and bases can be combined, influencing the proton-transfer process and the final ionicity of the PILs, consequently affecting their lubricating ability. Guo et al. [93] studied the influence of the ionicity and extent of hydrogen bonding of three PILs under steel-steel and steel-aluminum contact as neat lubricants and as additives to a mineral oil. As neat lubricants and 1 wt.% additives to a mineral oil, all three PILs studied reduced friction compared to the mineral oil in both tribological pairs. In general, the PIL with the highest ionicity and lower extent of hydrogen bonding showed the lowest friction reduction as a neat lubricant and the highest friction reduction as an additive to the mineral oil. The authors discussed that the potential cause of the higher friction as a neat lubricant may be due to the higher ionicity of this PIL, resulting in intense tribocorrosion reactions with the metal surfaces, particularly in aluminum. However, when used as only a 1 wt.% additive to the mineral oil, it is able to form a stable tribofilm in the absence of tribocorrosion reactions.

Recently, significant attention has been given to the use of PILs as water-based lubricant additives [94, 95] for different applications, such as in metalworking, micromachining, and biomedical lubrication. Moreover, some authors claim [92] that superlubricity can be achieved with water-based lubricants additized with imidazole-phosphate PILs.

3.5 Bio-based ILs

Environmental consciousness has seen a steady rise in recent years. As a result, there is a growing need to develop not only eco-friendly lubricants but also to recognize the importance of sourcing them from natural resources. Bio-compatible ILs, bio-based ILs, or more recently, ILs from natural resources are dominating the majority of research related to ILs lately [96–99]. Fatty acid-derived ILs [100, 101], amino acid-based ILs [76, 97], or Choline-based ILs [102, 103] have been proven to be good candidates for environment-friendly lubricant and additives. Choline amino acid ionic liquids have stood out among all the possibilities for having both the cation and the anion derived from natural sources. In Li et al. [104], a group of choline amino acid ionic liquids were synthesized using anions, and their lubricating ability was studied under steel-steel contact. In general, these choline amino acid ionic liquids demonstrated good friction-reducing and antiwear properties due to the formation of a physically absorbed film first and, in some cases, a tribolayer. In any case, lubricating ability is influenced by the anionic structure and the lubrication regime [104].

4. The nanoscale lubrication of ILs

In addition to macroscale lubrication, the study of ILs as thin films gains more and more attention from tribologists. Micro and nanoelectromechanical systems (MEMS/NEMS) as one of the most significant technologies have possessed critical applications in sensing, signal processing, wireless communications, automobiles, and medical devices. The large surface area to volume ratio of the moving parts in MEMS/NEMS make them vulnerable to adhesion, friction, and wear with inadequate lubrication, therefore, the advanced nanometer-thick lubricants are critical to the reliability of the devices. The ideal lubricant candidates are expected to have load-carrying, self-healing capability, and high thermal stability at interfaces [105].

Compared to the conventional thin lubrication films, such as PFPEs, IL thin films have demonstrated their potential in lubricating micro and nanoelectromechanical systems (MEMS/NEMS) and hard disk drives (HDDs) due to their flexibility in molecular designs, excellent thermal property and electrical conductivity, and adaptability to multiple applications [106]. Although the investigations of ILs as lubricant thin films are gradually increasing where the molecular structures and the chain length of cations and anions as well as the surface chemical characteristics of the substrates have been found to influence their tribological properties, but the major studies are still about APILs instead of PILs. Meanwhile, the understanding of their lubricating mechanisms and the impact of the environmental factors (such as humidity, normal pressure, sliding velocity, surface roughness, temperature, etc.) on their tribological performance is still limited.

4.1 The dynamic lubrication of IL thin films

The dynamic behaviors of ILs under confinement or at interfaces have been widely studied. MEMS/NEMS nanodevices are mainly designed or fabricated based on silicon material which is not inherently lubricious. The study of ILs to lubricate silicon surfaces was first reported by Yu et al. [107] in 2006. Imidazolium based ILs with different functional groups in cations and hexafluorophosphate anion are spin-coated as ultra-thin films on single-crystal silicon wafers with different terminals.

The vinyl-group functionalized IL films were found to reduce friction and wear of both the hydroxyl-terminated and vinyl group modified silicon surfaces with a ball-on-plate test configuration. Vinyl-group functionalized IL film also exhibited lower friction and better wear resistance than the commercial lube PFPEs (Adol-2000 in this case) in terms of the hydroxyl-terminated wafer. Good flexibility of IL molecules under confinement and the strong hydrogen bond and/or physico-chemical adsorptions at the interface of IL films and substrate contribute to the excellent tribological performance. In the study of Zhu et al. [108], three dip-coated imidazolium-based IL thin films with different anions, tetrafluoroborate, hexafluorophosphate, and adipate are evaluated on hydroxyl-terminated and amino-terminated Si substrates using a UMT-2MT tribometer. The hydrophilic properties of the IL films are proposed to affect their lubrication where the IL with hydrophobic hexafluorophosphate exhibits the best tribological performance and longer film durability at a severe condition.

In addition to the microscale tribological evaluations, Bhushan's group [109, 110] applied atomic force microscopy (AFM) to characterize the nanoscale adhesion, friction, and wear behaviors of imidazolium based ILs with different types of anions (hexafluorophosphate and octyl sulfate) on silicon substrates. The partially bonded films with a fraction of mobile lube present the lowest friction on the Si surface compared to fully bonded films and untreated films at microscale and nanoscale evaluations. At the nanoscale, the two imidazolium-based ILs present a slightly higher friction coefficient than the control lube PFPEs (Z-TETRAOL in this case) on the silica substrate. The IL with octyl sulfate anion causes more wear than the one with hexafluorophosphate anion and Z-TETRAOL. While from the microscale measurements with a ball-on-flat tribometer, both partially bonded IL films showed enhanced tribological performance with respect to Z-TETRAOL. It is noted that the friction coefficient value scale collected from ball-on-flat tribometer is much higher than that obtained from AFM. All the difference is attributed to the length scales of the test techniques and the applied normal load/pressure which may lead to different lubrication mechanisms. Zhao et al. [111] studied the microscale and nanoscale tribological properties of tetrafluoroborate-based IL films with different cations. The flexibility and surface energy of IL films are pointed to dominate the friction behavior at nanoscale while the rigid cycle structure and flexible chain of ILs play a critical role on friction at microscale.

Most of the aforementioned studies are based on silicon wafer, the lubricity of IL thin films on other solid substrates, including mica [112, 113], diamond-like carbon [114], boron carbon nitride [115], aluminum [116], steel [117], titanium [77], and nickel alloys [60] have also been evaluated. The nanoscale friction response of six PILs is studied by Sweeney et al. [118] with silica colloidal probe friction force microscopy (FFM) on mica surface. For all PILs, as shown in **Figure 3**, multiple ion layers are formed between the probe and the mica surface when the normal load is low. Oppositely, a single ion layer (in the boundary layer regime) with a smaller friction coefficient is found under high normal load. The PIL with longer alkyl chains may lead to higher friction. It is attributed to the stronger solvophobic effect or additional hydrogen bonding groups in PILs which makes the attraction between ions in near surface layers stronger. It also points out that the interactions within and between ion layers are controllable by introducing steric hindrance or additional hydrogen bonding in PILs.

4.2 Interface reactions of IL thin film and substrates

The strong interactions of ILs with the solid surface make them outperform the conventional lubes [7]. The strength of the IL ions and substrate interaction is tunable by

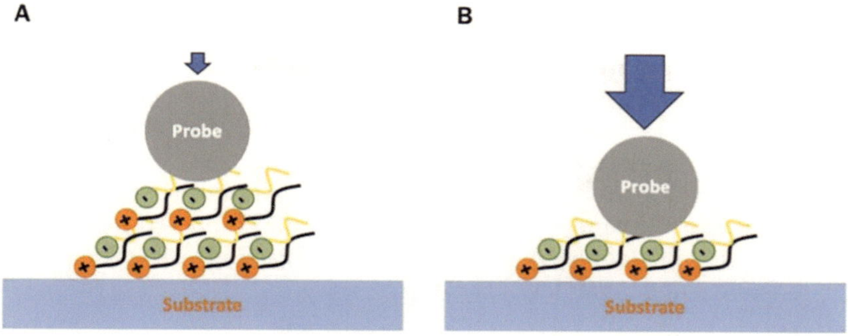

Figure 3.
Schematic diagram of (A) multiple ion layers formed under low normal load, and (B) single ion layer under high normal load.

varying the chemical structures of the IL ions that are both near the surface and in the subsequent layers [119–122], surface type [121, 123], and surface charge [119, 123, 124]. Specifically, for surface wetting and the nanoscale friction behavior, it is of importance to understand the 3D lateral structure of the lamellar-like IL thin film near the substrate surface or within the IL ion layers [106, 125].

Zhang et al. [122] found that when confined to a few nanometers, the rheological behaviors of five imidazolium ILs in the study vary strongly as a function of the number of IL ion layers with a surface force apparatus (SFA). The ILs' viscosities increased by several orders of magnitude, and the ILs transited from Newtonian to viscoelastic fluid and to an elastic solid. The phenomenon is attributed to the confinement-induced formation of supramolecular clusters with long relaxation time. The higher friction is related to the strain relaxation which is by means of the diffusion of those supramolecular clusters. However, IL 1-ethyl-3-methylimidazolium ethylsulfate exhibits the smallest friction coefficient where its strong hydrogen bonds are hypothesized to restrict the reorganization of the confined IL ions into clusters and impede its viscoelastic/elastic behavior. In addition, under both dry and humid (relative humidity is less than 3%) conditions, hydrophilic IL is found to remain a Newtonian behavior providing the best lubrication, but the hydrophobic IL exhibits a solid-like behavior with a poor tribological performance. In contrast, in Mo's group [126, 127], four chloride-based IL thin films with different functional groups (methyl, hydroxyl, nitrile, and carboxyl group) in the imidazolium cations, and four 1-propyl-3-methylimidazolium IL films with different anions are prepared. Generally, the hydrophilic properties of the cationic and anionic groups in IL films will facilitate the formation of a meniscus and increase the adhesion between the tip and IL film. For example, in the first set of ILs, the 1-ethanol-3-methylimidazolium chloride with hydroxyl end group in IL is found to have the highest adhesive force on the hydroxyl terminated Si substrate at the humidity of 15%, and it also shows the higher friction. The different findings between the two research groups might be due to the characteristics of the substrates, where the former is mica surfaces and the latter is glass-silicon surfaces, the number of confined IL ions layers, and the relative humidity.

In addition to the IL thin films that are spin-coated/dip-coated directly on the substrates, self-assembled monolayers (SAMs) as bifunctional chemical linkers are examined in some studies to improve the tribological properties of IL thin films [106, 125, 128]. SAMs due to their high packing density, low power consumption, and solid-like properties have been studied as lubricant films and show good load-carrying property but poor durability because of low molecular mobility [129, 130]. Pu et al. [128] found that octadecyltrichlorosilane (OTS) modified silicon surface improved the IL film quality with better wettability. The dual-layer films showed lower adhesive force and nominal friction coefficient than their corresponding IL film. The good adhesion resistance and better lubricity of the dual-layer indicate that a proper amount of mobile lubricants with low shear strength at the ILs- octadecyltrichlorosilane (OTS) interface is needed to maintain low nano-friction. Gusain et al. [106] proposed similar synergic effect of solid-like SAM and flowable IL fraction which exhibits better friction-reducing, wear-resistivity, and better load-carrying capacity.

5. Conclusions

ILs have obtained significant attention as novel lubricants and lubricant additives due to their unique properties. Their high thermal stability and conductivity make them ideal for harsh working conditions where traditional lubricants are not suitable. Another remarkable advantage of ILs is their highly tunable structure. By selecting specific combinations of cations and anions, researchers are able to tailor the physico-chemical properties of ILs to meet the specific requirements of different tribological applications. This tunability is vital for the development of novel and optimized lubricant solutions.

However, there still remain many challenges. Several promising future directions might unlock the full potential of ILs in lubrication and are summarized below:

- Developing predictive models: creating robust models that accurately predict the properties of IL based on their structure and combination. Meanwhile, combining AI models with traditional tribology to predict ILs' tribological performance and facilitate targeted design and optimization.

- Exploring novel IL structures: investigating task-specific ionic liquids and other innovative designs to achieve desired functionalities and improve the adaptive capacity to various operating environments.

- Nanotechnology integration: utilizing the potential of nanomaterials like nanoparticles and graphene to enhance key properties like thermal conductivity and tribofilm formation.

- ILs as additives in thin films: ILs have high capacity to self-assemble in a amphiphile self-assembly solvent which opens a door for them to be used as additives to low-cost thin film lubricants.

- Sustainability considerations: focusing on developing biodegradable or low toxic ILs and exploring cost-effective production methods to ensure environmental and economic sustainability.

Conflict of interest

The authors declare no conflict of interest.

DOI: http://dx.doi.org/10.5772/intechopen.1005538

Acronyms and abbreviations

IL	ionic liquids
APIL	aprotic ionic liquids
PIL	protic ionic liquids
$[NTf_2]$	bis(trifluoromethylsulfonyl)amide
$[C_{12}MIM]$	1-dodecyl-3-methylimidazolium
$[N_{4441}]$	tributylmethylammonium
$[N_{8881}]$	methyltrioctylammonium
$[C_6mim]$	1-hexyl-3-methylimidazolium
[Cl]	chloride
$[C_2mim]$	1-ethyl-3-methylimidazolium
OECD	Organization for Economic Co-operation and Development
ASTM	American Society for Testing and Materials
ISO	International Organization for Standardization
COD	chemical oxygen demand
BOD5	biochemical oxygen demand
PFPE	perfluoropolyether

Author details

Hong Guo[1,2*], Junru Pang[3] and Patricia Iglesias[4]

1 Hardware Development Engineering, HDD Business, Western Digital, USA

2 Electrical Engineering and Computer Science, Pennsylvania State University, USA

3 Mechanical Engineering Department, Pennsylvania State University, USA

4 Mechanical Engineering Department, Kate Gleason College of Engineering, Rochester Institute of Technology Rochester, USA

*Address all correspondence to: hxg6557@rit.edu

References

[1] Wilkes JS, Levisky JA, Wilson RA, Hussey CL. Dialkylimidazolium chloroaluminate melts: A new class of room-temperature ionic liquids for electrochemistry, spectroscopy and synthesis. Inorganic Chemistry. 1982;**21**:1263-1264. DOI: 10.1021/ic00133a078

[2] Lei Z, Chen B, Koo Y-M, MacFarlane DR. Introduction: Ionic liquids. Chemical Reviews. 2017;**117**:6633-6635. DOI: 10.1021/acs.chemrev.7b00246

[3] Hussain SMS, Adewunmi AA, Alade OS, Murtaza M, Mahboob A, Khan HJ, et al. A review of ionic liquids: Recent synthetic advances and oilfield applications. Journal of the Taiwan Institute of Chemical Engineers. 2023;**153**:105195. DOI: 10.1016/j.jtice.2023.105195

[4] Fajardo OY, Bresme F, Kornyshev AA, Urbakh M. Electrotunable lubricity with ionic liquid nanoscale films. Scientific Reports. 2015;**5**:7698. DOI: 10.1038/srep07698

[5] Minami I. Ionic liquids in tribology. Molecules. 2009;**14**:2286-2305. DOI: 10.3390/molecules14062286

[6] Waheed S, Ahmed A, Abid M, Mufti RA, Ferreira F, Bashir MN, et al. Ionic liquids as lubricants: An overview of recent developments. Journal of Molecular Structure. 2023;**1301**:137307

[7] Hayes R, Warr GG, Atkin R. Structure and nanostructure in ionic liquids. Chemical Reviews. 2015;**115**:6357-6426. DOI: 10.1021/cr500411q

[8] Ye C, Liu W, Chen Y, Yu L. Room-temperature ionic liquids: A novel versatile lubricant. Chemical Communications. 2001;(21):2244-2245

[9] Ye C, Liu W, Chen Y, Ou Z. Tribological behavior of Dy–sialon ceramics sliding against Si3N4 under lubrication of fluorine-containing oils. Wear. 2002;**253**:579-584

[10] Iglesias P, Bermúdez MD, Carrión FJ, Martnez-Nicolás G. Friction and wear of aluminium–steel contacts lubricated with ordered fluids-neutral and ionic liquid crystals as oil additives. Wear. 2004;**256**:386-392

[11] Qu J, Truhan JJ, Dai S, Luo H, Blau PJ. Ionic liquids with ammonium cations as lubricants or additives. Tribology Letters. 2006;**22**:207-214

[12] Minami I, Watanabe N, Nanao H, Mori S, Fukumoto K, Ohno H. Improvement in the tribological properties of imidazolium-derived ionic liquids by additive technology. Journal of Synthetic Lubrication. 2008;**25**:45-55

[13] Victoria PI. Tribological Properties of Aluminium-Based Materials. Spain: Universidad Politecnica de Cartagena; 2003

[14] Minami I, Kamimura H, Mori S. Thermo-oxidative stability of ionic liquids as lubricating fluids. Journal of Synthetic Lubrication. 2007;**24**:135-147

[15] Kamimura H, Kubo T, Minami I, Mori S. Effect and mechanism of additives for ionic liquids as new lubricants. Tribology International. 2007;**40**:620-625

[16] Phillips BS, John G, Zabinski JS. Surface chemistry of fluorine containing

ionic liquids on steel substrates at elevated temperature using Mössbauer spectroscopy. Tribology Letters. 2007;**26**:85-91

[17] Liu X, Zhou F, Liang Y, Liu W. Benzotriazole as the additive for ionic liquid lubricant: One pathway towards actual application of ionic liquids. Tribology Letters. 2006;**23**:191-196

[18] Guo H. Protic Ionic Liquids as Neat Lubricants & Lubricant Additives to Nonpolar and Polar Base Oils: Lubricating and Wear Mechanisms of Sliding Steel-Steel and Aluminum-Steel Contacts. ProQuest Dissertations & Theses. Rochester, NY: Rochester Institute of Technology; 2021

[19] Guo H, Iglesias P. Tribological behavior of ammonium-based protic ionic liquid as lubricant additive. Friction. 2021;**9**:169-178

[20] Sharma V, Doerr N, Aswath PB. Chemical–mechanical properties of tribofilms and their relationship to ionic liquid chemistry. RSC Advances. 2016;**6**:22341-22356

[21] Liu T. Understanding Structure-Property Relationships between Tribological Properties and Chemistry of Phosphonium Ionic Liquids. University of California, Merced ProQuest Dissertations & Theses. UC Merced; 2023

[22] Kreivaitis R, Treinytė J, Kupčinskas A, Gumbytė M, Andriušis A. Improving tribological properties of water/glycerol lubricating fluid by the synergy of nanoparticles and protic ionic liquid. Wear. 2023;**534**:205133

[23] Krick BA, Vail JR, Persson BNJ, Sawyer WG. Optical in situ micro tribometer for analysis of real contact area for contact mechanics, adhesion, and sliding experiments. Tribology Letters. 2012;**45**:185-194

[24] Pivnic K, Bresme F, Kornyshev AA, Urbakh M. Electrotunable friction in diluted room temperature ionic liquids: Implications for nanotribology. ACS Applied Nano Materials. 2020;**3**:10708-10719

[25] Khonsari MM, Booser ER. Applied Tribology: Bearing Design and Lubrication. Hoboken, NJ, USA: John Wiley & Sons; 2017

[26] Silva W, Zanatta M, Ferreira AS, Corvo MC, Cabrita EJ. Revisiting ionic liquid structure-property relationship: A critical analysis. International Journal of Molecular Sciences. 2020;**21**:7745

[27] Bapat AP, Erck R, Seymour BT, Zhao B, Cosimbescu L. Lipophilic polymethacrylate ionic liquids as lubricant additives. European Polymer Journal. 2018;**108**:38-47

[28] Song J. Research progress of ionic liquids as lubricants. ACS Omega. 2021;**6**:29345-29349

[29] Zhang S. Encyclopedia of Ionic Liquids. New York, NY: Springer Nature; 2023

[30] Blanco D, Oulego P, Ramos D, Fernández B, Cuetos JM. Model-free kinetics applied to evaluate the long-term thermal stability of three [NTf2] anion-based ionic liquids. Thermochimica Acta. 2017;**656**:70-84

[31] Liaw H-J, Chen K-Y, Chen H-Y, Liu S-N. Effect of heating temperature on the flash point of ionic liquids. Procedia Engineering. 2014;**84**:293-296

[32] Parajo JJ, Villanueva M, Troncoso J, Salgado J. Thermophysical properties of choline and pyridinium based ionic

liquids as advanced materials for energy applications. The Journal of Chemical Thermodynamics. 2020;**141**:105947

[33] Pereiro AB, Legido JL, Rodr A, et al. Physical properties of ionic liquids based on 1-alkyl-3-methylimidazolium cation and hexafluorophosphate as anion and temperature dependence. The Journal of Chemical Thermodynamics. 2007;**39**:1168-1175

[34] Vila J, Ginés P, Pico JM, Franjo C, Jiménez E, Varela LM, et al. Temperature dependence of the electrical conductivity in EMIM-based ionic liquids: Evidence of Vogel–Tamman–Fulcher behavior. Fluid Phase Equilibria. 2006;**242**:141-146

[35] Boruń A, Fernandez C, Bald A. Conductance studies of aqueous ionic liquids solutions [emim][BF4] and [bmim][BF4] at temperatures from (283.15 to 318.15) K. International Journal of Electrochemical Science. 2015;**10**:2120-2129

[36] Guo H, Adukure AR, Iglesias P. Effect of ionicity of three protic ionic liquids as neat lubricants and lubricant additives to a biolubricant. Coatings. 2019;**9**:713

[37] Wijanarko W, Khanmohammadi H, Espallargas N. Ionic liquids as boundary additives in water-based and PAO lubricants. Friction. 2022;**10**:1405-1423

[38] Avilés M-D, Pamies R, Sanes J, Bermúdez M-D. Graphene-ionic liquid thin film nanolubricant. Nanomaterials. 2020;**10**:535

[39] Vyavhare K, Aswath PB. Tribological properties of novel multi-walled carbon nanotubes and phosphorus containing ionic liquid hybrids in grease. Frontiers of Mechanical Engineering. 2019;**5**:15

[40] Murphy T, Varela LM, Webber GB, Warr GG, Atkin R. Nanostructure–thermal conductivity relationships in protic ionic liquids. The Journal of Physical Chemistry. B. 2014;**118**:12017-12024

[41] Rodil E, Arce A Jr, Arce A, Soto A. Measurements of the density, refractive index, electrical conductivity, thermal conductivity and dynamic viscosity for tributylmethylphosphonium and methylsulfate based ionic liquids. Thermochimica Acta. 2018;**664**:81-90

[42] Fröba AP, Rausch MH, Krzeminski K, Assenbaum D, Wasserscheid P, Leipertz A. Thermal conductivity of ionic liquids: Measurement and prediction. International Journal of Thermophysics. 2010;**31**:2059-2077

[43] Arkhipova EA, Ivanov AS, Maslakov KI, Savilov SV, Lunin VV. Effect of cation structure of tetraalkylammonium-and imidazolium-based ionic liquids on their conductivity. Electrochimica Acta. 2019;**297**:842-849

[44] Harjani JR, Farrell J, Garcia MT, Singer RD, Scammells PJ. Further investigation of the biodegradability of imidazolium ionic liquids. Green Chemistry. 2009;**11**:821-829

[45] Bubalo MC, Radošević K, Redovniković IR, Halambek J, Srček VG. A brief overview of the potential environmental hazards of ionic liquids. Ecotoxicology and Environmental Safety. 2014;**99**:1-12

[46] Quijano G, Couvert A, Amrane A, Darracq G, Couriol C, Le Cloirec P, et al. Toxicity and biodegradability of ionic liquids: New perspectives towards whole-cell biotechnological applications. Chemical Engineering Journal. 2011;**174**:27-32

[47] Steudte S, Bemowsky S, Mahrova M, Bottin-Weber U, Tojo-Suarez E,

Stepnowski P, et al. Toxicity and biodegradability of dicationic ionic liquids. RSC Advances. 2014;**4**:5198-5205

[48] Harjani JR, Singer RD, Garcia MT, Scammells PJ. Biodegradable pyridinium ionic liquids: Design, synthesis and evaluation. Green Chemistry. 2009;**11**:83-90

[49] Viesca JL, Oulego P, González R, Guo H, Battez AH, Iglesias P. Miscibility, corrosion and environmental properties of six hexanoate-and sulfonate-based protic ionic liquids. Journal of Molecular Liquids. 2021;**322**:114561

[50] Oulego P, Blanco D, Ramos D, Viesca JL, Díaz M, Hernández BA. Environmental properties of phosphonium, imidazolium and ammonium cation-based ionic liquids as potential lubricant additives. Journal of Molecular Liquids. 2018;**272**:937-947. DOI: 10.1016/j.molliq.2018.10.106

[51] Wan R, Xia X, Wang P, Huo W, Dong H, Chang Z. Toxicity of imidazoles ionic liquid [C16mim] Cl to HepG2 cells. Toxicology In Vitro. 2018;**52**:1-7

[52] Romero A, Santos A, Tojo J, Rodríguez A. Toxicity and biodegradability of imidazolium ionic liquids. Journal of Hazardous Materials. 2008;**151**:268-273

[53] EC. Regulation (EC) No 1907/2006 of the European Parliament and of the Council Concerning the Registration, Evaluation, Authorisation and Restriction of Chemicals (REACH), Establishing a European Chemicals Agency. Helsinki, Finland: European Agency for Safety and Health at Work; 2006

[54] Montalbán MG, Villora G, Licence P. Ecotoxicity assessment of dicationic versus monocationic ionic liquids as a more environmentally friendly alternative. Ecotoxicology and Environmental Safety. 2018;**150**:129-135

[55] Peric B, Sierra J, Martí E, Cruañas R, Garau MA. A comparative study of the terrestrial ecotoxicity of selected protic and aprotic ionic liquids. Chemosphere. 2014;**108**:418-425

[56] Reid JESJ, Prydderch H, Spulak M, Shimizu S, Walker AJ, Gathergood N. Green profiling of aprotic versus protic ionic liquids: Synthesis and microbial toxicity of analogous structures. Sustainable Chemistry and Pharmacy. 2018;**7**:17-26

[57] Baaqel H, Hallett JP, Guillen-Gosalbez G, Chachuat B. Sustainability assessment of alternative synthesis routes to aprotic ionic liquids: The case of 1-butyl-3-methylimidazolium tetrafluoroborate for fuel desulfurization. ACS Sustainable Chemistry & Engineering. 2021;**10**:323-331

[58] Jiménez AE, Bermúdez MD, Iglesias P, Carrión FJ, Martínez-Nicolás G. 1-N-alkyl-3-methylimidazolium ionic liquids as neat lubricants and lubricant additives in steel–aluminium contacts. Wear. 2006;**260**:766-782

[59] Mahrova M, Pagano F, Pejakovic V, Valea A, Kalin M, Igartua A, et al. Pyridinium based dicationic ionic liquids as base lubricants or lubricant additives. Tribology International. 2015;**82**:245-254

[60] Jiménez AE, Bermúdez MD, Iglesias P. Lubrication of Inconel 600 with ionic liquids at high temperature. Tribology International. 2009;**42**:1744-1751

[61] Jiménez AE, Bermúdez M-D. Ionic liquids as lubricants of titanium–steel contact. Tribology Letters. 2009;**33**:111-126

[62] Qu J, Bansal DG, Yu B, Howe JY, Luo H, Dai S, et al. Antiwear performance and mechanism of an oil-miscible ionic liquid as a lubricant additive. ACS Applied Materials & Interfaces. 2012;**4**:997-1002

[63] Kondou Y, Koyama T, Sasaki S. Ionic liquids as novel lubricants. Journal of the Vacuum Society of Japan. 2013;**56**:77-82

[64] Jimenez A-E, Bermúdez M-D. Ionic liquids as lubricants for steel–aluminum contacts at low and elevated temperatures. Tribology Letters. 2007;**26**:53-60

[65] Lu Q, Wang H, Ye C, Liu W, Xue Q. Room temperature ionic liquid 1-ethyl-3-hexylimidazolium-bis (trifluoromethylsulfonyl)-imide as lubricant for steel–steel contact. Tribology International. 2004;**37**:547-552

[66] Liu W, Ye C, Chen Y, Ou Z, Sun DC. Tribological behavior of sialon ceramics sliding against steel lubricated by fluorine-containing oils. Tribology International. 2002;**35**:503-509

[67] Fox MF, Priest M. Tribological properties of ionic liquids as lubricants and additives. Part 1: Synergistic tribofilm formation between ionic liquids and tricresyl phosphate. Proceedings of the Institution of Mechanical Engineers, Part J: Journal of Engineering Tribology. 2008;**222**:291-303

[68] Jiménez AE, Bermúdez MD, Carrión FJ, Martínez-Nicolás G. Room temperature ionic liquids as lubricant additives in steel–aluminium contacts: Influence of sliding velocity, normal load and temperature. Wear. 2006;**261**:347-359

[69] Kronberger M, Pejaković V, Gabler C, Kalin M. How anion and cation species influence the tribology of a green lubricant based on ionic liquids. Proceedings of the Institution of Mechanical Engineers, Part J: Journal of Engineering Tribology. 2012;**226**:933-951

[70] Zhou Y, Dyck J, Graham TW, Luo H, Leonard DN, Qu J. Ionic liquids composed of phosphonium cations and organophosphate, carboxylate, and sulfonate anions as lubricant antiwear additives. Langmuir. 2014;**30**:13301-13311

[71] Zhu L, Dong J, Zeng Q, Chao M, Li W, Gong K, et al. Study of two green phosphonium-based ionic liquids lubricants for steel/steel contact. Lubrication Science. 2022;**34**:467-479

[72] Khanmohammadi H, Wijanarko W, Espallargas N. The role of tribofilm chemical composition on wear of austenitic stainless steel lubricated with water-glycol containing ionic-liquids as additives. Wear. 2023;**520**:204672

[73] Liu X, Zhou F, Liang Y, Liu W. Tribological performance of phosphonium based ionic liquids for an aluminum-on-steel system and opinions on lubrication mechanism. Wear. 2006;**261**:1174-1179

[74] Kreivaitis R, Kupčinskas A, Žunda A, Ta TN, Horng JH. Effect of temperature on the lubrication ability of two ammonium ionic liquids. Wear. 2022;**492**:204217

[75] Dong R, Wen P, Zhang S, Zhang C, Sun W, Fan M, et al. The synthesis and tribological properties of dicarboxylic acid ionic liquids. Tribology International. 2017;**114**:132-140

[76] Zhu L, Dong J, Zeng Q, Chao M, Gong K, Li W, et al. A comprehensive study of amino acids based ionic liquids as green lubricants for various contacts. Tribology International. 2021;**162**:107137

[77] Jiménez AE, Bermúdez MD. Ionic liquids as lubricants of titanium–steel contact. Part 2: Friction, wear and surface interactions at high temperature. Tribology Letters. 2010;**37**:431-443

[78] Sanes MJ. Ionic liquids as lubricants of metal-polymer contacts. Preparation and properties of the first dispersions of ionic liquids and nanoparticles in polymers [PhD thesis]. ProQuest Dissertations & Theses. Spain: Universidad Politecnica de Cartagena; 2008

[79] Cai M, Liang Y, Zhou F, Liu W. Ionic liquids as lubricants. In: Green Solvents II. Dordrecht: Springer Netherlands; 2012. pp. 203-233. DOI: 10.1007/978-94-007-2891-2_9

[80] Yao M, Liang Y, Xia Y, Zhou F. Bisimidazolium ionic liquids as the high-performance antiwear additives in poly (ethylene glycol) for steel-steel contacts. ACS Applied Materials & Interfaces. 2009;**1**:467-471

[81] Qu J, Luo H, Chi M, Ma C, Blau PJ, Dai S, et al. Comparison of an oil-miscible ionic liquid and ZDDP as a lubricant anti-wear additive. Tribology International. 2014;**71**:88-97

[82] Barnhill WC, Qu J, Luo H, Meyer HM III, Ma C, Chi M, et al. Phosphonium-organophosphate ionic liquids as lubricant additives: Effects of cation structure on physicochemical and tribological characteristics. ACS Applied Materials & Interfaces. 2014;**6**:22585-22593

[83] Minami I, Inada T, Okada Y. Tribological properties of halogen-free ionic liquids. Proceedings of the Institution of Mechanical Engineers, Part J: Journal of Engineering Tribology. 2012;**226**:891-902

[84] Kondo Y, Yagi S, Koyama T, Tsuboi R, Sasaki S. Lubricity and corrosiveness of ionic liquids for steel-on-steel sliding contacts. Proceedings of the Institution of Mechanical Engineers, Part J: Journal of Engineering Tribology. 2012;**226**:991-1006

[85] Totolin V, Minami I, Gabler C, Dörr N. Halogen-free borate ionic liquids as novel lubricants for tribological applications. Tribology International. 2013;**67**:191-198

[86] Al Kaisy GMJ, Mutalib MIA, Rao T, Senatore A. Tribological performance of low viscosity halogen-free ammonium based protic ionic liquids with carboxylate anions as neat lubricants. Tribology International. 2021;**160**:107058

[87] Zhang C, Li F, Yang Z, Wang X, Chen G, Lu Z, et al. Comparing tribology properties of halogen-free ionic liquid, halogen-containing ionic liquid, and PAO 10 lubricants for steel–Al2024 friction contact at room temperature and high temperature. Journal of Molecular Liquids. 2021;**323**:115041

[88] Zhu L, Dong J, Ma Y, Jia Y, Peng C, Li W, et al. Synthesis and investigation of halogen-free phosphonium-based ionic liquids for lubrication applications. Tribology Transactions. 2019;**62**:943-954

[89] Saurín N, Minami I, Sanes J, Bermúdez MD. Study of the effect of tribo-materials and surface finish on the lubricant performance of new halogen-free room temperature ionic liquids. Applied Surface Science. 2016;**366**:464-474

[90] Greaves TL, Drummond CJ. Protic ionic liquids: Properties and applications. Chemical Reviews. 2008;**108**:206-237

[91] Guo H, Iglesias VP. Ionic liquids as high-performance lubricants and lubricant additives. In: Ionic

Liquids—Thermophysical Properties and Applications. London, UK: IntechOpen; 2021

[92] Zheng Z, Xu Y, Chen H, Liu X, Wang H, Meng W, et al. Simple But Effective Liquid Superlubricity Triggered by Protic Ionic Liquids Aqueous. Available at SSRN 4668460. n.d

[93] Guo H, Pang J, Adukure AR, Iglesias P. Influence of hydrogen bonding and ionicity of protic ionic liquids on lubricating steel–steel and steel–aluminum contacts: Potential ecofriendly lubricants and additives. Tribology Letters. 2020;**68**:1-10

[94] Kaneko K, Akamatsu M, Sakai K, Sakai H. Lubrication by adsorption films of hydrophilic amine-based protic ionic liquids: Effect of anion species. Journal of Oleo Science. 2021;**70**:1615-1621

[95] de Castro VV, dos Santos LM, Antonini LM, Schroeder RM, Mattedi S, Souza KS, et al. Water-based lubricant containing protic ionic liquids and talc lubricant particles: Wear and corrosion analysis. Wear. 2023;**518**:204633

[96] Dås AK, Sequeira RA, Maity TK, Prasad K. Bio-ionic liquid promoted selective coagulation of κ-carrageenan from *Kappaphycus alvarezii* extract. Food Hydrocolloids. 2021;**111**:106382

[97] Dou M, Feng Q, Yang J, Zou S, Li K, Li Z, et al. Evaluation of rheological properties and tribological properties of choline/acetylamino acid ionic liquids at different frequencies. Wear. 2024;**540**:205247

[98] Avilés MD, Pamies R, Sanes J, Arias-Pardilla J, Carrión FJ, Bermúdez MD. Protic ammonium bio-based ionic liquid crystal lubricants. Tribology International. 2021;**158**:106917

[99] Ruan X, Chen H, Jiao C, Wei X, Wang Y, Gong P, et al. Simultaneous low friction and ultra-low wear enabled by a 2,5-furandicarboxylic acid derived bio-based ionic liquid. Tribology International. 2022;**174**:107683

[100] Fan M, Ma L, Zhang C, Wang Z, Ruan J, Han M, et al. Biobased green lubricants: Physicochemical, tribological and toxicological properties of fatty acid ionic liquids. Tribology Transactions. 2018;**61**:195-206

[101] Gusain R, Khan A, Khatri OP. Fatty acid-derived ionic liquids as renewable lubricant additives: Effect of chain length and unsaturation. Journal of Molecular Liquids. 2020;**301**:112322

[102] Zheng D, Wang X, Zhang M, Ju C. Synergistic effects between the two choline-based ionic liquids as lubricant additives in glycerol aqueous solution. Tribology Letters. 2019;**67**:1-13

[103] Gontrani L. Choline-amino acid ionic liquids: Past and recent achievements about the structure and properties of these really "green" chemicals. Biophysical Reviews. 2018;**10**:873-880

[104] Li Y, Yang F, Li Y, Cai M, Li H, Fan X, et al. Choline amino acid ionic Liquids: A novel green potential lubricant. Journal of Molecular Liquids. 2022;**360**:119539

[105] Gong X, Li L. Nanometer-thick ionic liquids as boundary lubricants. Advanced Engineering Materials. 2018;**20**(5). DOI: 10.1002/adem.201700617

[106] Gusain R, Kokufu S, Bakshi PS, Utsunomiya T, Ichii T, Sugimura H, et al. Self-assembled thin film of imidazolium ionic liquid on a silicon surface: Low friction and remarkable wear-resistivity. Applied Surface Science.

2016;**364**:878-885. DOI: 10.1016/j.apsusc.2015.12.175

[107] Yu B, Zhou F, Mu Z, Liang Y, Liu W. Tribological properties of ultra-thin ionic liquid films on single-crystal silicon wafers with functionalized surfaces. Tribology International. 2006;**39**:879-887. DOI: 10.1016/j.triboint.2005.07.039

[108] Zhu M, Yan J, Mo Y, Bai M. Effect of the anion on the tribological properties of ionic liquid nano-films on surface-modified silicon wafers. Tribology Letters. 2008;**29**:177-183. DOI: 10.1007/s11249-007-9294-x

[109] Bhushan B, Palacio M, Kinzig B. AFM-based nanotribological and electrical characterization of ultrathin wear-resistant ionic liquid films. Journal of Colloid and Interface Science. 2008;**317**:275-287. DOI: 10.1016/j.jcis.2007.09.046

[110] Palacio M, Bhushan B. Ultrathin wear-resistant ionic liquid films for novel MEMS/NEMS applications. Advanced Materials. 2008;**20**:1194-1198. DOI: 10.1002/adma.200702006

[111] Zhao W, Mo Y, Pu J, Bai M. Effect of cation on micro/nano-tribological properties of ultra-thin ionic liquid films. Tribology International. 2009;**42**:828-835. DOI: 10.1016/j.triboint.2008.10.020

[112] Perkin S, Albrecht T, Klein J. Layering and shear properties of an ionic liquid, 1-ethyl-3-methylimidazolium ethylsulfate, confined to nano-films between mica surfaces. Physical Chemistry Chemical Physics. 2010;**12**:1243-1247. DOI: 10.1039/B920571C

[113] Smith AM, Lovelock KRJ, Gosvami NN, Welton T, Perkin S. Quantized friction across ionic liquid thin films. Physical Chemistry Chemical Physics. 2013;**15**:15317. DOI: 10.1039/c3cp52779d

[114] Sasaki S. Environmentally friendly tribology (eco-tribology). Journal of Mechanical Science and Technology. 2010;**24**:67-71. DOI: 10.1007/s12206-009-1154-1

[115] Chen YM, Zeng ZX, Yang SR, Zhang JY. The tribological performance of BCN films under ionic liquids lubrication. Diamond and Related Materials. 2009;**18**:20-26. DOI: 10.1016/j.diamond.2008.07.023

[116] Qu J, Chi M, Meyer HM, Blau PJ, Dai S, Luo H. Nanostructure and composition of tribo-boundary films formed in ionic liquid lubrication. Tribology Letters. 2011;**43**:205-211. DOI: 10.1007/s11249-011-9800-z

[117] Lawes SDA, Hainsworth SV, Blake P, Ryder KS, Abbott AP. Lubrication of steel/steel contacts by choline chloride ionic liquids. Tribology Letters. 2010;**37**:103-110. DOI: 10.1007/s11249-009-9495-6

[118] Sweeney J, Webber GB, Rutland MW, Atkin R. Effect of ion structure on nanoscale friction in protic ionic liquids. Physical Chemistry Chemical Physics. 2014;**16**:16651. DOI: 10.1039/C4CP02320J

[119] Li H, Endres F, Atkin R. Effect of alkyl chain length and anion species on the interfacial nanostructure of ionic liquids at the Au(111)–ionic liquid interface as a function of potential. Physical Chemistry Chemical Physics. 2013;**15**:14624. DOI: 10.1039/c3cp52421c

[120] Hayes R, Warr GG, Atkin R. At the interface: Solvation and designing ionic liquids. Physical Chemistry Chemical Physics. 2010;**12**:1709. DOI: 10.1039/b920393a

[121] Di Lecce S, Kornyshev AA, Urbakh M, Bresme F. Structural effects in nanotribology of nanoscale films of ionic liquids confined between metallic surfaces. Physical Chemistry Chemical Physics. 2021;**23**:22174-22183. DOI: 10.1039/D1CP03345J

[122] Zhang X, Han M, Espinosa-Marzal RM. Thin-film rheology and tribology of imidazolium ionic liquids. ACS Applied Materials & Interfaces. 2023;**15**:45485-45497. DOI: 10.1021/acsami.3c10018

[123] Li H, Wood RJ, Endres F, Atkin R. Influence of alkyl chain length and anion species on ionic liquid structure at the graphite interface as a function of applied potential. Journal of Physics: Condensed Matter. 2014;**26**:284115. DOI: 10.1088/0953-8984/26/28/284115

[124] Hayes R, Borisenko N, Tam MK, Howlett PC, Endres F, Atkin R. Double layer structure of ionic liquids at the Au(111) electrode Interface: An atomic force microscopy investigation. The Journal of Physical Chemistry C. 2011;**115**:6855-6863. DOI: 10.1021/jp200544b

[125] Liu J, Li J, Yu B, Ma B, Zhu Y, Song X, et al. Tribological properties of self-assembled monolayers of catecholic imidazolium and the spin-coated films of ionic liquids. Langmuir. 2011;**27**:11324-11331. DOI: 10.1021/la201378b

[126] Mo Y, Zhao W, Zhu M, Bai M. Nano/microtribological properties of ultrathin functionalized imidazolium wear-resistant ionic liquid films on single crystal silicon. Tribology Letters. 2008;**32**:143-151. DOI: 10.1007/s11249-008-9371-9

[127] Mo Y, Huang F, Zhao F. Functionalized imidazolium wear-resistant ionic liquid ultrathin films for MEMS/NEMS applications. Surface and Interface Analysis. 2011;**43**:1006-1014. DOI: 10.1002/sia.3684

[128] Pu J, Jiang D, Mo Y, Wang L, Xue Q. Micro/nano-tribological behaviors of crown-type phosphate ionic liquid ultrathin films on self-assembled monolayer modified silicon. Surface and Coating Technology. 2011;**205**:4855-4863. DOI: 10.1016/j.surfcoat.2011.04.089

[129] Spearing SM. Materials issues in microelectromechanical systems (MEMS). Acta Materialia. 2000;**48**:179-196. DOI: 10.1016/S1359-6454(99)00294-3

[130] Tsukruk VV. Molecular lubricants and glues for micro- and nanodevices. Advanced Materials. 2001;**13**:95-108. DOI: 10.1002/1521-4095 (200101) 13:2<95::AID-ADMA95>3. 0.CO;2-J